# SpringerBriefs in Mathematics of Planet Earth • Weather, Climate, Oceans



**Managing Series Editors**

Dan Crisan, Imperial College London, London, UK
Darryl Holm, Imperial College London, London, UK

**Series Editors**

Colin Cotter, Imperial College London, London, UK
Jochen Broecker, University of Reading, Reading, UK
Ted Shepherd, University of Reading, Reading, UK
Sebastian Reich, University Potsdam, Potsdam, Germany
Valerio Lucarini, University Hamburg, Hamburg, Germany

SpringerBriefs present concise summaries of cutting-edge research and practical applications across a wide spectrum of fields. Featuring compact volumes of 50 to 125 pages, the series covers a range of content from professional to academic. Briefs are characterized by fast, global electronic dissemination, standard publishing contracts, standardized manuscript preparation and formatting guidelines, and expedited production schedules.

Typical topics might include:

- A timely report of state-of-the art techniques
- A bridge between new research results, as published in journal articles, and a contextual literature review
- A snapshot of a hot or emerging topic
- An in-depth case study

SpringerBriefs in the Mathematics of Planet Earth showcase topics of current relevance to the Mathematics of Planet Earth. Published titles will feature both academic-inspired work and more practitioner-oriented material, with a focus on the application of recent mathematical advances from the fields of Stochastic And Deterministic Evolution Equations, Dynamical Systems, Data Assimilation, Numerical Analysis, Probability and Statistics, Computational Methods to areas such as climate prediction, numerical weather forecasting at global and regional scales, multi-scale modelling of coupled ocean-atmosphere dynamics, adaptation, mitigation and resilience to climate change, etc. This series is intended for mathematicians and other scientists with interest in the Mathematics of Planet Earth.

More information about this subseries at http://www.springer.com/series/15250

Thomas H. Gibson · Andrew T. T. McRae ·
Colin J. Cotter · Lawrence Mitchell ·
David A. Ham

# Compatible Finite Element Methods for Geophysical Flows

## Automation and Implementation Using Firedrake

Springer

Thomas H. Gibson
Department of Mathematics
Imperial College London
London, UK

Colin J. Cotter
Department of Mathematics
Imperial College London
London, UK

David A. Ham
Department of Mathematics
Imperial College London
London, UK

Andrew T. T. McRae
Department of Physics
Oxford University
Oxford, UK

Lawrence Mitchell
Department of Computer Science
Durham University
Durham, UK

SpringerBriefs in Mathematics of Planet Earth - Weather, Climate, Oceans
ISSN 2509-7326 ISSN 2509-7334 (electronic)
ISBN 978-3-030-23956-5 ISBN 978-3-030-23957-2 (eBook)
https://doi.org/10.1007/978-3-030-23957-2

Mathematics Subject Classification (2010): 97N80, 74S05, 58D30



This Springer imprint is published by the registered company Springer Nature Switzerland AG
The registered company address is: Gewerbestrasse 11, 6330 Cham, Switzerland

# Preface

The development of simulation software is an important aspect of modern scientific computing, especially in the geosciences. Developing complex numerical code requires a large time investment and a range of knowledge spanning several academic disciplines. Arriving at a physical description of a complex physical system, such as a coupled atmosphere, ocean, and land model, demands acute awareness of domain sciences: meteorology, oceanography, biochemistry, ecology, and rheology. Discretising the governing partial differential equations to produce a stable numerical scheme requires expertise in mathematical analysis, and its translation into efficient code for massively parallel systems demands advanced knowledge in low-level code optimisation and computer architectures. Therefore, development of such software is a multidisciplinary effort and its design must enable scientists across several disciplines to collaborate effectively.

Software projects involving automatic code generation have become increasingly popular in recent years, as these help create a separation of concerns between different aspects of development. This allows for agile collaboration between computer scientists with expertise in hardware and software, computational scientists with expertise in numerical algorithms, and domain scientists such as meteorologists, oceanographers, and climate scientists.

The finite element method is a standard mathematical framework for computing numerical solutions of partial differential equations. It has been widely used in engineering applications for many decades, due to its ease of use on unstructured grids and in complicated geometries. It has become increasingly popular in fluids and solids models within geosciences, and its formulation is highly amenable to code-generation techniques. A weak formulation of the relevant PDEs, together with a mesh and appropriate discrete function spaces, is enough to characterise the problem completely.

The models we present in this book use 'compatible', or 'mimetic', finite element discretisations. While their use in fluids problems is relatively new, compared to the more standard continuous and discontinuous Galerkin methods, their use in other applications can be traced back to the late 1970s. In a geophysical context, these discretisations are a generalisation of the C-grid horizontal variable staggering to a finite element setting. They therefore inherit the C-grid's properties such as good representation of near-grid-scale waves.

The topics in this book are non-exhaustive; the purpose of this text is to provide the reader with an idea for how one might use Firedrake as a base for their own numerical codes. The book is organised in the following way. To establish context, the reader is provided with a gentle introduction to modeling geophysical flows in Chap. 1. This includes summaries of common model hierarchies and desirable properties of numerical models for operational use. Chapter 2 presents the application of finite element modeling for two- and three-dimensional systems. Full discretisations are presented using the framework of compatible finite element methods. Using the Firedrake finite element library is a central aspect of this book, and therefore Chap. 3 is devoted to introduce the reader to basic concepts and examples needed to implement the discretisations summarised in Chap. 2. Both Chaps. 4 and 5 provide numerical implementations of examples using Firedrake in two and three dimensions respectively. In particular, Chap. 5 discusses efficient algorithms for solving the resulting discrete systems, which are essential for real-world operational settings.

London, UK — Thomas H. Gibson
Oxford, UK — Andrew T. T. McRae
London, UK — Colin J. Cotter
Durham, UK — Lawrence Mitchell
London, UK — David A. Ham
April 2019

**Acknowledgements** This work was supported by the Engineering and Physical Sciences Research Council (grant numbers EP/L016613/1, EP/M011054/1, EP/L000407/1, and EP/R029628/1), the Natural Environment Research Council (grant numbers NE/R008795/1 and NE/K008951/1), and the European Research Council (ERC) under the European Union's Horizon 2020 research and innovation programme (grant agreement no. 741112).

# Contents

# Chapter 1
# Geophysical Fluid Dynamics and Simulation

In this chapter we describe various models for the atmosphere; they will be discretised in the rest of the book. We start by considering models with the fewest approximations and work our way down to simplified models. Model hierarchies have formed the backbone of geophysical fluid dynamics: simpler models are more tractable and easier to analyse, whilst more complex models bring us closer to modelling the real physical system. Moving up and down the hierarchy allows us to trace physical phenomena to mathematical calculations on simpler models. Since the inception of numerical computer simulation, model hierarchies have also been useful as a way to trade off model accuracy with computational cost.

## 1.1 Hierarchies of Models

The first computer weather forecast on the ENIAC (one of the world's first digital computers) by Charney, von Neumann and coworkers solved a single layer quasi-geostrophic model[1] (Lynch, 2008). Subsequently, as computer power increased, it became possible to move up the hierarchy, making fewer model approximations in more numerically intensive calculations. Numerical weather predictions were made with multilayer shallow water models, hydrostatic compressible Euler models, and finally the non-hydrostatic compressible Euler models which represent the state of the art today (Kalnay, 2003; Bauer et al., 2015; Staniforth and Wood, 2008). Model hierarchies are also very useful in the development of numerical models. One can move down the hierarchy to isolate specific aspects of the model to examine their numerical treatment in a less computationally intensive setting, and then move back up to use this insight. In the development of atmospheric dynamical cores, it is very

[1] We will not cover the quasi-geostrophic model in this book. It is an approximation that filters out acoustic and internal gravity waves, hence allowing a larger timestep, which made the computation feasible on the ENIAC.



T. H. Gibson et al., *Compatible Finite Element Methods for Geophysical Flows*, Mathematics of Planet Earth, https://doi.org/10.1007/978-3-030-23957-2_1

standard to start with the shallow water equations to focus on horizontal discretisation aspects before moving up the hierarchy.

The hierarchy of models we will build up in this chapter is shown in Figure 1.1. This is just one choice of hierarchy, since for example one can apply the hydrostatic approximation directly to the compressible Euler equations without making Boussinesq/anelastic approximations. Similarly, we do not consider quasi-geostrophic approximations which are valid for rapidly rotating systems; these approximations can be made at any step of our hierarchy.

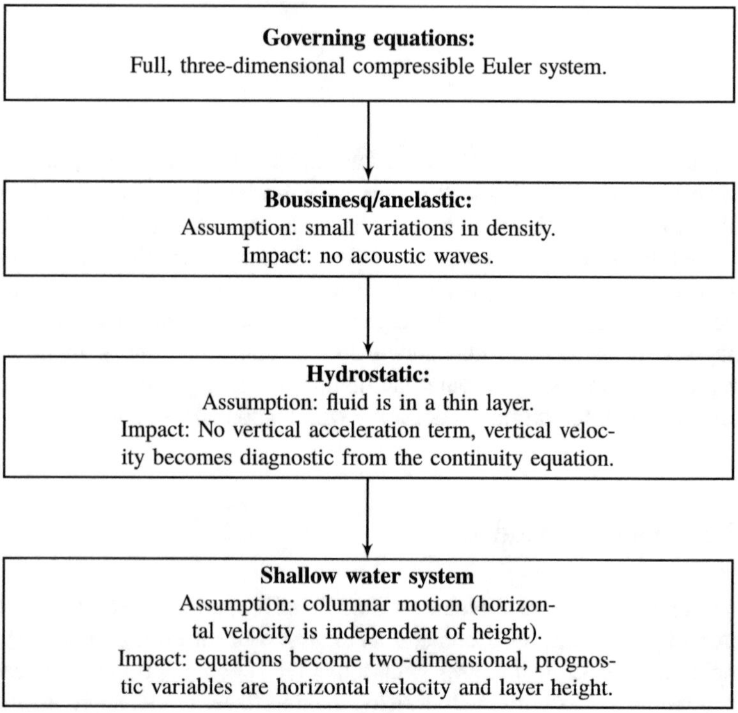


**Fig. 1.1** A hierarchy of models illustrating a successive application of various approximations to yield simplified models

### 1.1.1 The Compressible Euler System

We start by presenting the compressible Euler equations, which have the fewest approximations amongst our hierarchy. We assume that the air is dry (no moisture), inviscid (no viscous forces), and adiabatic (no sources or diffusion of temperature). The governing equations for a dry, inviscid, adiabatic, compressible fluid in a rotating reference frame with angular velocity $\mathbf{\Omega}$ may be written in the form

$$\boxed{\begin{aligned} &\frac{D\boldsymbol{u}}{Dt} + 2\boldsymbol{\Omega}\times\boldsymbol{u} = -\frac{1}{\rho}\nabla p - \nabla\Phi, &(1.1)\\ &\frac{\partial\rho}{\partial t} + \nabla\cdot(\boldsymbol{u}\rho) = 0, &(1.2)\\ &\frac{D\theta}{Dt} = 0, &(1.3)\\ &p = P(\rho,T), &(1.4)\end{aligned}}$$

where: $\boldsymbol{u}$ is the fluid velocity, $\rho$ is the fluid density, $p$ is the pressure, and $\Phi$ is the geopotential comprising the gravitational and centrifugal potentials (often neglected as it is small compared to the gravitational potential). We use the potential temperature $\theta$, defined as the temperature an air parcel would attain if moved adiabatically to a reference pressure $p_R$. The ideal gas laws allow us to compute this explicitly:

$$\frac{T}{p^\kappa} = \frac{\theta}{p_R^\kappa} \implies \theta = T\left(\frac{p_R}{p}\right)^\kappa, \tag{1.5}$$

where $p_R$ is a chosen reference pressure, and $\kappa = R/c_p$ is the ratio of the gas constant $R$ and the specific heat at constant pressure $c_p$. The payoff for this rather complicated formulation is that the potential temperature is constant along Lagrangian trajectories in the absence of diabatic processes.

Given some velocity field $\boldsymbol{u}$, we denote the material derivative by $\frac{D}{Dt}$. For a field $F$, $\frac{DF}{Dt}$ is given by the instantaneous time rate of change of $F$, plus a contribution from the spatial variation of $F$ as a result of being moved by $\boldsymbol{u}$,

$$\frac{DF}{Dt} \equiv \frac{\partial F}{\partial t} + (\boldsymbol{u}\cdot\nabla)F. \tag{1.6}$$

If the material derivative is zero, then the field is materially conserved following the motion of the fluid. Thus we interpret the above equations as material derivatives of various fluid quantities, especially after noticing that

$$0 = \frac{\partial\rho}{\partial t} + \nabla\cdot(\boldsymbol{u}\rho) \equiv \frac{D\rho}{Dt} + \rho\nabla\cdot\boldsymbol{u}. \tag{1.7}$$

Equation (1.4) closes the system of equations by relating pressure to the other thermodynamic variables. In the case of the atmosphere, the equation of state for an ideal gas is typically used, i.e.,

$$P(\rho,T) = \rho RT, \tag{1.8}$$

where $R$ is the gas constant for dry air.

It is fairly common to use an alternative formulation of the pressure gradient term

$$\frac{1}{\rho}\nabla p = c_p\theta\nabla\Pi, \tag{1.9}$$

where $\Pi$ is the Exner pressure given by the equation of state

$$\Pi = \left(\frac{R\rho\theta}{p_r}\right)^{\frac{\kappa}{1-\kappa}}. \tag{1.10}$$

One of the reasons for making this change is that it is then fairly simple to incorporate the thermodynamic effects of moisture into the pressure gradient term. The situation is more complicated in the case of the ocean, where the equation of state additionally depends on salinity. In this book we shall concentrate on the dry adiabatic case, but the techniques developed here are directly extensible to model the moist atmosphere, and the ocean.

These equations also need boundary conditions. In this book we shall mostly restrict ourselves to slip boundary conditions $\boldsymbol{u} \cdot \boldsymbol{n} = 0$ where $\boldsymbol{n}$ is the vector normal to the boundary. Another important boundary condition is the free surface boundary condition, $p = p_A$ (where $p_A$ is an external reference pressure). In this case, the boundary surface must move with the velocity $\boldsymbol{u}$ at the boundary.

We note for later that (1.2) is an expression of local mass conservation; integration over a control volume $V$ leads to

$$\frac{\mathrm{d}}{\mathrm{d}t}\int_V \rho \,\mathrm{d}\boldsymbol{x} + \int_{\partial V} \boldsymbol{u} \cdot \boldsymbol{n} \,\mathrm{d}S = 0, \tag{1.11}$$

where $\boldsymbol{n}$ is the outward pointing unit normal vector to the boundary $\partial V$ of $V$. This shows that the rate of change of total mass in $V$ is balanced by fluxes through the boundary $\partial V$.

At this point, it is worth discussing various approximations to the Coriolis term $2\boldsymbol{\Omega} \times \boldsymbol{u}$. For a model on the sphere, $\boldsymbol{\Omega}$ is aligned with the polar axis. A common approximation is the *traditional approximation*, under which the vertical part of this force is neglected (since it is small compared to the gravitational force). For a coordinate system whose origin is at the centre of the sphere, we write $\hat{\boldsymbol{r}} = \boldsymbol{x}/|\boldsymbol{x}|$ for the unit vector pointing away from the origin. Then the traditional approximation replaces $\boldsymbol{\Omega}$ with $f\hat{\boldsymbol{r}}$, where $f = \boldsymbol{\Omega} \cdot \hat{\boldsymbol{r}}$. Under this approximation, the Coriolis term vanishes at the equator and is maximum at the poles. We call $f$ the Coriolis parameter.

For both mathematical simplicity and in the study of various models, a patch of the planetary surface can further be approximated by a plane that is tangent at one point. We consider two common treatments of the Coriolis parameter:

1. *$f$-plane approximation*: $f \equiv f_0$ is taken to be constant-valued. This approximation is used frequently in the case of highly idealised flows. A notable consequence is that Rossby waves, which depend on variations in $f$, do not occur in models using this approximation.
2. *$\beta$-plane approximation*: Since $f$ depends on variations in latitude, an $f$-plane approximation may not be appropriate when considering flows over large length scales. The $\beta$-plane approximation improves on this by considering leading-order variations: $f = f(y) = f_0 + \beta y$.

### 1.1.2 Anelastic and Boussinesq Approximations

The next level down in our model hierarchy is distinguished by the lack of acoustic waves that are present in the compressible Euler model. The anelastic approximation was first introduced in Ogura and Phillips (1962) to filter out acoustic modes without needing to assume hydrostatic balance. It has since been well-studied and modified (Durran, 1989, 2008). The approximation is based upon the assumption that there are only small variations in density relative to a reference field. We begin by writing

$$\rho = \tilde{\rho} + \delta\rho(\boldsymbol{x}, t), \tag{1.12}$$

where $\tilde{\rho}$ is a reference density field that only depends on the height (i.e., the radial direction on the sphere or the $z$ direction in the plane).

After noticing that $\frac{\partial \tilde{\rho}}{\partial t} = 0$, we rewrite (1.2) as

$$\frac{\partial \delta\rho}{\partial t} + \nabla \cdot (\boldsymbol{u}\tilde{\rho}) + \nabla \cdot (\boldsymbol{u}\delta\rho) = 0. \tag{1.13}$$

Since $\delta\rho$ is much smaller than $\tilde{\rho}$, we neglect those terms to obtain the anelastic continuity equation

$$\nabla \cdot (\tilde{\rho}\boldsymbol{u}) = 0. \tag{1.14}$$

We now consider the pressure gradient term under this approximation:

$$p = \tilde{p} + \delta p(\boldsymbol{x}, t), \tag{1.15}$$

where $\tilde{p}$ is chosen such that it is the hydrostatic pressure corresponding to $\tilde{\rho}$, i.e., satisfying

$$\hat{\boldsymbol{k}} \cdot \nabla \tilde{p} = -g\tilde{\rho}, \tag{1.16}$$

where $\hat{\boldsymbol{k}}$ is the unit vector in the up direction (either $\hat{\boldsymbol{r}}$ for the sphere or $\hat{z}$ for a planar geometry). After this choice, the combination of the pressure gradient and the gravity force becomes

$$-\nabla p - g\hat{\boldsymbol{k}}\rho = -\nabla\delta p - g\delta\rho\hat{\boldsymbol{k}}. \tag{1.17}$$

The momentum equation (1.1) is then

$$\frac{D\boldsymbol{u}}{Dt} + 2\boldsymbol{\Omega} \times \boldsymbol{u} = -\frac{1}{\rho}\nabla\delta p - g\frac{\delta\rho}{\rho}. \tag{1.18}$$

The anelastic approximation first replaces $\rho$ with $\tilde{\rho}$ everywhere whilst leaving $\delta\rho$ in the gravity term, and we get

$$\frac{D\boldsymbol{u}}{Dt} + 2\boldsymbol{\Omega} \times \boldsymbol{u} = -\frac{1}{\tilde{\rho}}\nabla\delta p - g\frac{\delta\rho}{\tilde{\rho}}. \tag{1.19}$$

In the case where $\tilde{\rho}$ is constant, say $\rho \equiv \rho_0$, we obtain the Boussinesq equations. These are relevant in ocean modelling, since the density in the ocean does not deviate far from a constant reference value $\rho_0$.

The complete set of Boussinesq equations are

$$\boxed{\begin{aligned} \frac{D\boldsymbol{u}}{Dt} + 2\boldsymbol{\Omega} \times \boldsymbol{u} &= -\nabla \bar{p} - b\hat{z}, && (1.20)\\ \nabla \cdot \boldsymbol{u} &= 0, && (1.21)\\ \frac{D\theta}{Dt} &= 0, && (1.22)\\ b &= B(\theta), && (1.23) \end{aligned}}$$

where $\bar{p} = \delta p/\rho_0$, and $B$ is an equation of state either derived from (1.4), or more frequently, a linear approximation $B(\theta) = B_0 + \alpha\theta$, where $B_0$ and $\alpha$ are constants. Ocean models also contain a salinity, $S$, which is materially advected,

$$\frac{DS}{Dt} = 0. \tag{1.24}$$

The buoyancy also depends on this salinity, resulting in a more general equation of state $B(\theta, S, z)$ that also includes mean pressure variations in height. There is no known closed form of this equation and it must be approximated by fitting approximations to data (Jackett and McDougall, 1997).

Another useful modification leads to the compressible Boussinesq equations, in which (1.21) is replaced by

$$\frac{\partial p}{\partial t} + c^2 \nabla \cdot \boldsymbol{u} = 0, \tag{1.25}$$

where $c$ is an acoustic wave speed. Taking the limit $c \to \infty$ recovers the standard Boussinesq equations. This modification allows the introduction of compressibility effects without the full nonlinear pressure gradient term of the compressible Euler equations. This is particularly useful for testing implicit solver approaches in a simplified setting.

To complete the anelastic approximation for non-constant $\tilde{\rho}$, we turn our attention to just the vertical part of (1.19). With $z$ as the coordinate in the vertical direction, we have

$$\frac{Dw}{Dt} = -\frac{1}{\tilde{\rho}}\frac{\partial \delta p}{\partial z} - g\frac{\delta\rho}{\tilde{\rho}} = -\frac{\partial}{\partial z}\left(\frac{\delta p}{\tilde{\rho}}\right) - \frac{\delta p}{\tilde{\rho}}\frac{\partial \ln(\tilde{\rho})}{\partial z} - g\frac{\delta\rho}{\tilde{\rho}}, \tag{1.26}$$

where the latter equality comes from the chain rule. Recalling the definition of potential temperature in (1.5), it can be shown using the ideal gas law that

$$c_p \ln(\theta) = c_v \ln(p) - c_p \ln(\rho) - \ln(R) - \frac{R}{c_p}\ln(p_R), \tag{1.27}$$

implying that

$$\frac{\delta\theta}{\theta} \approx \frac{1}{\gamma}\frac{\delta p}{\tilde{p}} - \frac{\delta\rho}{\tilde{\rho}}, \tag{1.28}$$

where $\gamma = c_p/c_v$ and $\delta\theta$ denotes variations in potential temperature (Vallis, 2017, Sect. 2.5). We can now eliminate $\delta\rho$ in the gravitational term of (1.26), using (1.27) and (1.28) to get

$$\begin{aligned}\frac{Dw}{Dt} &\approx g\left(\frac{\delta\theta}{\theta} - \frac{1}{\gamma}\frac{\delta p}{\tilde{p}}\right) - \frac{\partial}{\partial z}\left(\frac{\delta p}{\tilde{\rho}}\right) + \frac{\partial}{\partial z}\left(\ln(\theta) - \frac{1}{\gamma}\ln(\tilde{p})\right)\frac{\delta p}{\tilde{\rho}} \\ &= g\frac{\delta\theta}{\theta} - \frac{\partial}{\partial z}\left(\frac{\delta p}{\tilde{\rho}}\right) + \frac{\partial \ln(\theta)}{\partial z}\frac{\delta p}{\tilde{\rho}},\end{aligned} \tag{1.29}$$

where the final equality in (1.29) is derived by applying (1.16). We now make a final observation by recognising that the term $\partial \ln(\theta)/\partial z = \theta^{-1}\partial\theta/\partial z$ actually defines a length scale in potential temperature. In the atmosphere, the vertical length scale in potential temperature (roughly 100 km) is much larger than the density vertical length scale (about 10 km), so the last term in (1.29) can be neglected under the anelastic approximation. This gives us a vertical momentum equation of the form

$$\frac{Dw}{Dt} = b_a - \frac{\partial \bar{p}}{\partial z}, \tag{1.30}$$

where $b_a = g\delta\theta/\theta$ is the anelastic buoyancy and $\bar{p} = \delta p/\tilde{\rho}$. The horizontal momentum equations simplify in a far more direct manner by simply neglecting density variations. The complete set of anelastic equations are given by the system

$$\frac{D\boldsymbol{u}}{Dt} + 2\boldsymbol{\Omega}\times\boldsymbol{u} = b_a\hat{z} - \nabla\bar{p}, \tag{1.31}$$

$$\nabla\cdot(\tilde{\rho}\boldsymbol{u}) = 0, \tag{1.32}$$

$$\frac{Db_a}{Dt} = 0. \tag{1.33}$$

These equations are similar in presentation to that of Ogura and Phillips (1962) and Vallis (2017). However, various forms of the equations exist (often referred to as pseudo-incompressible or modified-anelastic sets) such as the set proposed in Durran (1989).

Both the Boussinesq equations for the ocean and the anelastic equations for the atmosphere (together with the local Cartesian approximation) provide another level of models which approximate the more general Euler equations. They do not permit acoustic modes, and both have been very productive in limited applications, such as large-eddy or localised flow simulations. However, particularly for the atmosphere, they are not generally sufficient for global-scale circulation modelling. A more in-depth discussion of the relative merits of different approximations for global-scale modelling can be found in Davies et al. (2003).

### 1.1.3 Hydrostatic Approximation

The hydrostatic approximation is valid under the assumption of a thin layer of fluid, i.e., the equations are being solved in a domain of horizontal scale $L$ and vertical scale $H$, with $H \ll L$. We assume that the vertical velocity has a typical scale $W$ whilst the horizontal velocity has a typical scale $U$. We write the advection equation (1.33) in the form

$$\frac{D_h b}{Dt} + wN^2 = 0, \tag{1.34}$$

where $\frac{D_h b}{Dt} = b_t + \boldsymbol{u}_H \cdot \nabla b$ and $\boldsymbol{u}_H$ is the horizontal component of the velocity, and where $N^2 = \frac{\partial b}{\partial z}$. This then implies that

$$W \approx \frac{bU}{LN^2}. \tag{1.35}$$

Hence, $\frac{Dw}{Dt} \ll b$ if

$$\frac{U^2}{L^2N^2} = \frac{\epsilon^2}{\mathrm{Ri}}, \tag{1.36}$$

where $\epsilon = H/L$ is the aspect ratio of the domain, and $\mathrm{Ri} = N^2H^2/U^2$ is the Richardson number of the flow. This means that we can neglect the vertical component of the acceleration $\frac{Du}{Dt}$ if the aspect ratio $\epsilon$ is small or the Richardson number Ri is large.

In the hydrostatic approximation, we simply neglect the vertical acceleration:

$$\frac{D\boldsymbol{u}_H}{Dt} + 2\boldsymbol{\Omega} \times \boldsymbol{u} = b_a\hat{\boldsymbol{z}} - \nabla\bar{p}, \tag{1.37}$$

$$\nabla \cdot (\tilde{\rho}\boldsymbol{u}) = 0, \tag{1.38}$$

$$\frac{Db_a}{Dt} = 0. \tag{1.39}$$

As an aside, taking the Boussinesq equations (or, in the atmosphere, the anelastic equations) and applying the traditional and hydrostatic approximations yields the so-called *primitive equations*.

### 1.1.4 Single-Layer Rotating Shallow Water System

Thus far, we have consider three-dimensional models. Since the domains we are interested in geophysical modelling typically have high aspect ratios (the atmosphere is tens of kilometres high, but thousands of kilometres across), significant insight into large scale behaviour can be obtained by considering vertically-integrated models.

One of the simplest such models is that provided by the single-layer rotating shallow water equations. It allows for the analysis of flows under rotational effects within a relatively simple framework. It can be described as the culmination of all previous approximations mentioned in this section. That is, it describes a thin layer of constant density (akin to the Boussinesq/anelastic approximation) in hydrostatic balance. Under the shallow water approximation, we assume that the horizontal velocity is independent of depth, i.e., the fluid is moving as vertical columns. Thus, the system is inherently a two-dimensional approximation. The shallow water equations consider the case where the upper surface of the fluid is free, so the pressure is constant there, and the surface moves at the speed of the total velocity $(\boldsymbol{u}, w)$.

We start by summarising the continuity equation of a fluid with horizontal velocity $\boldsymbol{u}$ (which we previously denoted $\boldsymbol{u}_H$) and variable depth $h = h(x, y, t)$. This can be derived from three-dimensional mass conservation. Since the fluid has constant density, it is therefore incompressible and, in components, we have

$$\frac{\partial w}{\partial z} = -\left(\frac{\partial u}{\partial x} + \frac{\partial v}{\partial y}\right) \equiv -\nabla \cdot \boldsymbol{u}, \tag{1.40}$$

where $\nabla$ now denotes the horizontal gradient. If we consider a vertical column of fluid with bottom depth $z = \eta_b$ and top $z = \eta_t$, integrating the left-hand side of (1.40) and applying the fundamental theorem of calculus gives

$$\int_{\eta_b}^{\eta_t} \frac{\partial w}{\partial z}\,\mathrm{d}z = w(\eta_t) - w(\eta_b) = -h\nabla \cdot \boldsymbol{u}, \tag{1.41}$$

where $h(x, y, t) = \eta_t(x, y, t) - \eta_b(x, y, t)$. We assume that no mass is lost at the free surface $\eta_t$. That is, no fluid crosses the surface (which is analogous to the ocean). The height of the fluid parcel is therefore embedded in the free surface, therefore

$$\frac{D\eta_t}{Dt} = w(\eta_t). \tag{1.42}$$

Similarly, the motion of the fluid must follow the bottom topography:

$$\frac{D\eta_b}{Dt} = w(\eta_b). \tag{1.43}$$

Using (1.42) and (1.43), (1.41) becomes

$$\frac{D}{Dt}(\eta_t - \eta_b) \equiv \frac{Dh}{Dt} = -h\nabla \cdot \boldsymbol{u}, \tag{1.44}$$

resulting in the continuity equation for the shallow water system,

$$\frac{Dh}{Dt} + h\nabla \cdot \boldsymbol{u} \equiv \frac{\partial h}{\partial t} + \nabla \cdot (\boldsymbol{u}h) = 0. \tag{1.45}$$

This equation has the same form as (1.2), with a different (but related) interpretation. Here it describes conservation of the total volume of fluid beneath the free surface.

Momentum balance can be derived directly from the hydrostatic balance equation,

$$\frac{\partial p}{\partial z} = -\rho g, \tag{1.46}$$

and since $\rho$ is assumed to be constant, we may integrate directly to obtain

$$p(x, y, z, t) = -\rho g z + p_0. \tag{1.47}$$

At the top surface $z = \eta_t = h + \eta_b$, the pressure is determined by the weight of the fluid above, which is assumed to be negligible. Therefore we set $p(x, y, \eta_t) = 0$. The pressure inside the fluid layer is then

$$p(x, y, z, t) = \rho g(h(x, y, t) + \eta_b(x, y, t) - z). \tag{1.48}$$

Consequently, the horizontal pressure gradient is $z$-independent, with $\nabla p = \nabla(h + \eta_b)$. Therefore, the rotating single-layer shallow water equations are

$$\frac{D\boldsymbol{u}}{Dt} + f\boldsymbol{u}^{\perp} = -g\nabla(h + \eta_b), \tag{1.49}$$

$$\frac{\partial h}{\partial t} + \nabla \cdot (\boldsymbol{u}h) = 0. \tag{1.50}$$

We have introduced the notation $\boldsymbol{u}^{\perp} = \hat{\boldsymbol{z}} \times \boldsymbol{u}$, a rotation of the two-dimensional velocity $\boldsymbol{u}$ in the plane. No additional thermodynamic equations are needed to close this system of equations, as thermodynamic effects were eliminated in the various assumptions that led to this model.

To gain some further insight into these equations, we perform some further manipulations to (1.49)–(1.50). Introducing the relative vorticity $\zeta = \nabla^{\perp} \cdot \boldsymbol{u}$, we can rewrite the momentum equation in *vector-invariant* form:

$$\frac{\partial \boldsymbol{u}}{\partial t} + (\zeta + f)\boldsymbol{u}^{\perp} + \nabla\left(g(h + \eta_b) + \frac{1}{2}|\boldsymbol{u}|^2\right) = 0. \tag{1.51}$$

Defining the shallow water potential vorticity $q = \frac{\zeta + f}{h}$, this can be written as

$$\frac{\partial \boldsymbol{u}}{\partial t} + qh\boldsymbol{u}^{\perp} + \nabla\left(g(h + \eta_b) + \frac{1}{2}|\boldsymbol{u}|^2\right) = 0. \tag{1.52}$$

Taking the 2d-curl, and combining this with the shallow water continuity equation (1.50), one obtains an advection equation for potential vorticity:

$$\frac{\partial q}{\partial t} + (\boldsymbol{u} \cdot \nabla)q \equiv \frac{Dq}{Dt} = 0, \tag{1.53}$$

implying that $q$ remains constant and is materially conserved moving with the fluid velocity.

In a boundary-free domain, the shallow water equations have a conserved energy $E$, and mass $M$, given by

$$E = \frac{1}{2}\int_{\Omega} h|\boldsymbol{u}|^2 + g((h+\eta_b)^2 - \eta_b^2)\,\mathrm{d}\boldsymbol{x}, \tag{1.54}$$

$$M = \int_{\Omega} h\,\mathrm{d}\boldsymbol{x}. \tag{1.55}$$

In addition, there is an infinite hierarchy of conserved Casimirs based on potential vorticity:

$$C_i \equiv \int_{\Omega} q^i h\,\mathrm{d}\boldsymbol{x}, \qquad \text{for all } i = 1,2,\cdots. \tag{1.56}$$

Sometimes, it is useful to linearise the shallow water equations. In the absence of topography, a steady-state solution is a constant layer depth $H$ at rest. Small perturbations $\boldsymbol{u}$ and $h$ about this state then satisfy the linear equations

$$\frac{\partial \boldsymbol{u}}{\partial t} + f\boldsymbol{u}^{\perp} = -g\nabla h, \tag{1.57}$$

$$\frac{\partial h}{\partial t} + H\nabla\cdot\boldsymbol{u} = 0, \tag{1.58}$$

where $\boldsymbol{u}^{\perp} = \hat{z}\times\boldsymbol{u}$. It is easy to see that these equations conserve the linearised energy,

$$E = \frac{1}{2}\int H|\boldsymbol{u}|^2 + gh^2\,\mathrm{d}\boldsymbol{x}, \tag{1.59}$$

by direct calculation:

$$\begin{aligned}
\frac{\partial E}{\partial t} &= \int H\boldsymbol{u}\cdot\frac{\partial \boldsymbol{u}}{\partial t} + gh\frac{\partial h}{\partial t}\,\mathrm{d}\boldsymbol{x},\\
&= \int H\boldsymbol{u}\cdot(-g\nabla h - f\boldsymbol{u}^{\perp}) - ghH\nabla\cdot\boldsymbol{u}\,\mathrm{d}\boldsymbol{x},\\
&= \int -gH\boldsymbol{u}\cdot\nabla h - gHh\nabla\cdot\boldsymbol{u}\,\mathrm{d}\boldsymbol{x},\\
&= \int -gH\nabla\cdot(h\boldsymbol{u})\,\mathrm{d}\boldsymbol{x},\\
&= 0,
\end{aligned} \tag{1.60}$$

using the divergence theorem. The resulting surface term vanishes due to the boundary condition $\boldsymbol{u}\cdot\boldsymbol{n} = 0$.

These equations provide the simplest system that exhibits the slow-fast separation in geophysical fluid dynamics. To see this split, consider the $f$-plane approximation ($f \equiv f_0$, constant) with periodic boundary conditions. We use the Helmholtz decomposition $\boldsymbol{u} = \nabla^{\perp}\psi + \nabla\phi + \boldsymbol{u}_0$, where $\boldsymbol{u}_0$ is constant, and $\nabla^{\perp} = (-\partial_y, \partial_x)$, and where $\psi$ and $\phi$ have zero spatial mean. The linear equations become

$$\frac{\partial \psi}{\partial t} + f\phi = 0, \tag{1.61}$$

$$\frac{\partial \phi}{\partial t} - f\psi = -gh, \tag{1.62}$$

$$\frac{\partial h}{\partial t} + H\nabla^2\phi = 0, \tag{1.63}$$

$$\frac{\partial \boldsymbol{u}_0}{\partial t} + f\boldsymbol{u}_0^{\perp} = 0. \tag{1.64}$$

We notice that when $\psi = -gh/f$ and $\phi = 0$, we obtain a steady state solution. This solution corresponds to the state of geostrophic balance, $f\hat{z}\times\boldsymbol{u} = -g\nabla h$ which gives a divergence-free steady velocity. To find time-dependent solutions, we eliminate $\boldsymbol{u}$ and $\psi$ to obtain

$$\frac{\partial^2 \phi}{\partial t^2} + (f^2 - gH\nabla^2)\phi = 0. \tag{1.65}$$

This is a dispersive wave equation (unless $f = 0$ in which case all waves propagate at speed $\sqrt{gH}$). When $f \neq 0$, these waves are called inertia–gravity waves. This split between slow divergence-free solutions (which are, in fact, steady for these equations), and fast divergent waves is a critical aspect of large-scale balanced geophysical fluid dynamics, and it is important that numerical schemes can maintain this split at the discrete level. We also note that the constant part $\boldsymbol{u}_0$ (the *harmonic* part, which is both divergence- and curl-free), just oscillates with frequency $f$; this is called the inertial oscillation.

When $f = f_0 + \beta y$, this spatial variation in the Coriolis parameter introduces an additional mechanism that means that the divergence-free balanced solutions become slowly propagating. We assume that the solution is close to geostrophic balance, so that $\boldsymbol{u} = \boldsymbol{u}_g + \boldsymbol{u}_{ag}$ where $f_0\boldsymbol{u}_g^{\perp} = -g\nabla h$ (implying $\nabla\cdot\boldsymbol{u}_g = 0$, and $\boldsymbol{u}_g = \nabla^{\perp}\psi$ with $\psi = gh/f$). We also make the quasi-geostrophic approximation, assuming that $h$ is small compared to $H$. Then we get (at leading order)

$$\frac{\partial \boldsymbol{u}_g}{\partial t} + \beta y\boldsymbol{u}_g^{\perp} + f_0\boldsymbol{u}_{ag}^{\perp} = -g\nabla h, \tag{1.66}$$

$$\frac{f_0}{g}\frac{\partial \psi}{\partial t} + H\nabla\cdot\boldsymbol{u}_{ag} = 0. \tag{1.67}$$

The easiest way to understand this approximation is to take the curl of (1.66), to obtain

$$\frac{\partial}{\partial t}\nabla^2\psi + \beta\frac{\partial \psi}{\partial x} - f_0\nabla\cdot\boldsymbol{u}_{ag} = 0, \tag{1.68}$$

where we can finally use (1.67) to obtain

$$\frac{\partial}{\partial t}\left(\nabla^2 - \frac{f_0^2}{gH}\right)\psi + \beta\frac{\partial \psi}{\partial x} = 0. \tag{1.69}$$

Solutions of this equation are called Rossby waves, which are waves whose restoring force is the conservation of linear vorticity under $y$-variations in $f$. These are propagating divergence-free solutions that are slow compared to the unbalanced inertia–gravity waves. It is important that a numerical scheme preserves this separation into fast divergent and slow divergence-free components.

## 1.2 Numerical Modelling of Geophysical Flows

The use of computational models based on the numerical solution of PDEs to simulate physical processes is a powerful tool which complements experimentation, observation and theory. As the dynamics governing large-scale weather, ocean, and climate processes cannot be replicated in a laboratory, numerical simulation is often the only mode of scientific inquiry available to researchers aiming to test, build, and improve existing models. Here, we summarise some important developments in the computational models used in large-scale geophysical flows.

In contrast to typical engineering applications, where viscous and dissipative effects combined with outflow boundary conditions can alleviate the catastrophic accumulation of grid scale errors, large scale atmospheric modelling requires that the balanced forces of the continuous equations are exactly captured when discretising. If this is not the case, then numerical artefacts can easily swamp the true dynamics.

Structured orthogonal grids, such as the latitude–longitude grid, are among the most widely-used grids in global forecast and ocean circulation models. Problems associated with the latitude–longitude grid in a finite difference or finite volume ocean model have been discussed extensively as well as the desired numerical properties required of accurate simulation (Adcroft et al., 2004; Griffies et al., 2000; Higdon, 2006). Here, we will summarise these issues relevant in both atmosphere and ocean modelling.

On a full latitude–longitude grid, the convergence of the meridians leads to a clustering of resolution occurring at the poles. For an explicit time-integration method with an Eulerian advection scheme, the CFL condition (Courant et al., 1928) imposes unbearably small timesteps as the resolution increases. To circumvent this problem, semi-implicit time-integration combined with semi-Lagrangian advection schemes (for atmosphere models see Staniforth and Côté (1991), Williamson (2007); for ocean models Kumar Das andWeaver (1995)) are used to remove the severe timestep restriction. However semi-implicit time-integrators require the solution of a globally-defined elliptic PDE at every timestep. Significant data communication among the grid points at the poles is required and presents an inescapable computational bottleneck, as processor communication is slow and can leave other processors waiting for data (Staniforth and Thuburn, 2012).

The accurate and stable representation of large-scaled force balances in a finite difference model demands a staggered variable arrangement, with pressure and velocity unknowns stored at different locations on the grid. The canonical taxonomy of variable staggerings for finite difference models is given in Arakawa and Lamb (1977). In that nomenclature, finite difference atmosphere models almost exclusively use C-grid staggering, while ocean models tend towards either a B- or C-grid. The B-grid locates velocity at the cell centres and pressure at the vertices, while the C-grid locates pressure at the cell centres and the normal component of velocity at the cell facets. A comparison of both staggering schemes for implicit ocean models was performed by wubs et al. (2006). We will focus on the numerical properties of the C-grid.

In the context of horizontal discretisations, Staniforth and Thuburn (2012) considered a number of essential and desirable properties for an atmospheric dynamical core, which motivate the choice of compatible finite element discretisations discussed in this book. These properties are as follows:

1. Mass conservation of both active and passive fields. This is desirable, but not essential for short-term weather prediction, but is more important when considering long-term climate simulations.
2. Accurate representation of flow close to hydrostatic and geostrophic balance is crucial. Slowly evolving, balanced flows should be accurately represented upon discretisation; failure to do so results in a spontaneous loss of balance.
3. Computational modes should be absent, or at the very least well-controlled. Upon linearisation, both the continuous and discrete equations will support a spectrum of wave modes. Therefore, it is reasonable that the discrete wave modes will approximate those of the continuous equations. However, in some cases the discrete modes behave "unphysically" (zero frequency modes); we refer to these as "computational modes."
4. Grid imprinting should be minimal. Any non-uniform meshes will contain grid regions which are locally different to the vast majority of the mesh. This can lead to different error patterns, which may influence the resolved solution by coupling through nonlinearity. For example, the latitude–longitude grid has very different structure at the poles.
5. The geopotential term and pressure gradient should not produce unphysical sources of vorticity.
6. Terms involving the pressure should be energy conserving.
7. The discretisation of the Coriolis force should be energy conserving.
8. Rossby waves should not propagate unrealistically fast.
9. Axial angular momentum should be conserved.

Properties 5–9 all rely on the discretisation being "mimetic" in the sense that the discrete equations should mimic basic geometrical properties of the continuous system. These properties are all achievable with a finite difference C-grid staggering of the dynamic variables (Arakawa and Lamb, 1977), and hence this lays down the gauntlet for other discretisations.

There are two strong reasons why the latitude–longitude grid is used with the C-grid. The first is that the latitude–longitude grid is formed from quadrilaterals. This means that, globally, there are twice as many horizontal velocity values as pressure (each cell contains one pressure value, and four horizontal velocity values that are each shared with a neighbouring cell, so there are effectively two horizontal velocity values per cell). This mirrors the physical situation where each point has two horizontal velocity values and one pressure value, and helps to avoid spurious mode branches in the linear solution. Using a mesh constructed from other polygons alters the global ratio of velocity and pressure values. One possibility is to use triangular grids, which allow a pseudo-uniform coverage of the sphere from icosahedral triangulation, as used in the ICON model (Zängl et al., 2015), for example. With triangular cells, the shortage of velocity values (the ratio is 3:2 instead of 2:1) means that when $f = 0$ there are two branches of solutions to the inertia–gravity wave equation, one physical branch and one spurious branch. When $f \neq 0$, the situation is worse, with the two branches intertwined so that for some parts of the spectrum, one branch is the physical one, and for different parts of the spectrum, the other branch is. This causes serious problems when modelling baroclinic flows in three dimensions (Danilov, 2010; Gassmann, 2011. On hexagonal grids (the dual grid to the icosahedral triangulation is made up of hexagons plus 12 pentagons) there are more edges, and so there are an excess of velocity values (3:1 ratio). This excludes spurious inertia–gravity wave branches but there is now an extra spurious branch of Rossby waves (Thuburn, 2008). However, this extra branch is comprised of highly oscillatory solutions that have very slow phase velocity, which makes the solutions not much different from the high frequency Rossby waves which are also very slow. It was shown in Thuburn et al. (2014) that upwinding in velocity advection removes these spurious solutions very quickly, so they do not seem to be a concern for accurate modelling.

The second reason is that the latitude–longitude is used is that it is orthogonal. An orthogonal grid is one where the line joining the centres of two cells that share an edge crosses that edge at right-angles. This is a critical part of the classical C-grid formulation and was extended to arbitrary polygonal meshes in Ringler et al. (2010). The extension to non-orthogonal grids was provided in Thuburn and Cotter (2012).

To avoid the parallel scalability issues of the latitude–longitude grid, one possibility is to find other grids that cover the sphere in a more uniform manner. For the C-grid, three possibilities are the icosahedral grid (triangles), the dual-icosahedral grid (hexagons), and the cubed sphere made by tiling the faces of a cube and deforming to a sphere (quadrilaterals). The icosahedral grid has the spurious inertia–gravity waves mentioned above. The discretisation of the Coriolis term on the dual-icosahedral grid is inconsistent, *i.e.*, the discretisation does not become more accurate as the mesh is refined (Thuburn et al., 2014). The orthogonal version of the cubed sphere grid has a slightly less severe version of the problem of the latitude–longitude grid, in that the ratio of the largest and smallest cell areas goes to infinity as the mesh is refined, with clustering of grid points around the cube vertices (Putman and Lin, 2007). The non-orthogonal cubed sphere version of the C-grid was also demonstrated to be inconsistent in Thuburn et al. (2014).

One way to obtain an orthogonal mesh without resolution clustering at the poles is to use multiple overset grids, such as the so-called "Yin-Yang" or "tennis ball" mesh (for example, see Qaddouri and Lee 2011). However, coupling the solution between different grids in the overlap regions is challenging, and it seems that stability is dependent on keeping the width of the overlap region fixed as the mesh is refined, leading to communication overheads Staniforth and Thuburn (2012).

In contrast, compatible finite element methods present a way of simultaneously satisfying the properties listed above and avoiding polar singularities in the problem domain (Cotter and Shipton, 2012). The mimetic properties can be obtained without relying on an underlying orthogonal mesh structure. This allows for the use of far more general meshes, ranging from refined icosahedral sphere (triangular) and non-orthogonal cubed sphere (quadrilateral) meshes. Furthermore, the number of velocity and pressure degrees of freedom is less tightly coupled to the connectivity of the underlying mesh, and the finite element method presents a systematic mechanism for constructing higher order discretisations. The rest of this book is dedicated to explaining how to implement geophysical fluid dynamics models built around compatible finite element methods using Firedrake.

# Chapter 2
# Finite Element Methods for Geophysical Flows

In this chapter, we introduce some basic concepts for discretisation of geophysical fluid models, motivated by the compatible finite element framework. We begin by defining compatible finite element spaces in two dimensions, before using them to describe a compatible finite element discretisation of the shallow water equations. Then, we'll extend these ideas to three dimensions and discuss the spaces that are used to implement three dimensional geophysical flows with advected temperature.

## 2.1 Compatible Spaces in Two Dimensions

Compatible finite element spaces are important for geophysical fluid dynamics modelling because they preserve the Helmholtz decomposition of vector fields which describes the split into divergence-free and divergent fields. In this section we describe these structures in two dimensions.

We begin by defining some Hilbert spaces of functions of various types defined on a domain $\Omega \subset \mathbb{R}^2$, which provide the relevant setting for the Helmholtz decomposition. The $L^2(\Omega)$ norm of a scalar function $f$ is defined by

$$\|f\|_{L^2(\Omega)} = \sqrt{\int_\Omega f^2 \,\mathrm{d}\boldsymbol{x}}. \tag{2.1}$$

Then we define $L^2(\Omega)$ (abbreviated to $L^2$) as the space of functions with finite $L^2$ norm. Similarly, for vector fields $\boldsymbol{u}$ with (weak) divergence $\nabla \cdot \boldsymbol{u}$, the $H(\mathrm{div};\Omega)$ norm is

$$\|\boldsymbol{u}\|_{H(\mathrm{div};\Omega)} = \sqrt{\int_\Omega |\boldsymbol{u}|^2 + (\nabla \cdot \boldsymbol{u})^2 \,\mathrm{d}\boldsymbol{x}}. \tag{2.2}$$



T. H. Gibson et al., *Compatible Finite Element Methods for Geophysical Flows*, Mathematics of Planet Earth, https://doi.org/10.1007/978-3-030-23957-2_2

We define $H(\mathrm{div};\Omega)$ (abbreviated to $H(\mathrm{div})$) as the space of vector fields with finite $H(\mathrm{div};\Omega)$ norm. Finally, we define the $H^1(\Omega)$ norm (abbreviated to $H^1$) of a scalar function $f$ with (weak) gradient $\nabla f$ as

$$\|f\|_{H^1(\Omega)} = \sqrt{\int_\Omega f^2 + |\nabla f|^2 \,\mathrm{d}\boldsymbol{x}}. \tag{2.3}$$

These spaces are related by (weak) differential operators $\nabla^\perp$ and $\nabla\cdot$. Any vector field $\boldsymbol{u} \in H(\mathrm{div};\Omega)$ has a divergence $\nabla\cdot\boldsymbol{u} \in L^2(\Omega)$. Further, a function $f \in H^1(\Omega)$ has a well-defined gradient, so one can define a vector field $\boldsymbol{u} = \nabla^\perp f$ with vanishing divergence.

This structure comprises a de Rham complex on $\Omega$,

$$0 \rightarrow H^1(\Omega) \xrightarrow{\nabla^\perp} H(\mathrm{div};\Omega) \xrightarrow{\nabla\cdot} L^2(\Omega) \rightarrow 0, \tag{2.4}$$

which describes a sequence of spaces and operators between them, with $\nabla^\perp$ mapping onto the kernel of $\nabla\cdot$ in $H(\mathrm{div};\Omega)$, and $\nabla\cdot$ mapping onto the entire $L^2(\Omega)$.

We can also define dual operators:

$$\delta : L^2(\Omega) \rightarrow H(\mathrm{div};\Omega), \tag{2.5}$$

$$\delta^\perp : H(\mathrm{div};\Omega) \rightarrow H^1(\Omega), \tag{2.6}$$

which are characterised by the equations:

$$\int_\Omega \boldsymbol{w}\cdot\delta\phi \,\mathrm{d}\boldsymbol{x} = -\int_\Omega \nabla\cdot\boldsymbol{w}\phi\,\mathrm{d}\boldsymbol{x}, \quad \forall \boldsymbol{w} \in H(\mathrm{div};\Omega), \tag{2.7}$$

$$\int_\Omega \gamma\delta^\perp\cdot\boldsymbol{u}\,\mathrm{d}\boldsymbol{x} = -\int_\Omega \nabla^\perp\gamma\cdot\boldsymbol{u}\,\mathrm{d}\boldsymbol{x}, \quad \forall\gamma \in H^1(\Omega). \tag{2.8}$$

For periodic boundary conditions and smooth functions, $\delta$ and $\delta^\perp\cdot$ coincide with $\nabla$ and $\nabla^\perp\cdot$ respectively (because of the absence of boundary terms in the definition). The dual operators lead to the Helmholtz/Hodge decomposition of $H(\mathrm{div};\Omega)$,

$$H(\mathrm{div};\Omega) = \nabla^\perp H^1(\Omega) \oplus \mathfrak{h} \oplus \delta L^2(\Omega), \tag{2.9}$$

where

$$\nabla^\perp H^1(\Omega) = \left\{\nabla^\perp f : f \in H^1(\Omega)\right\}, \tag{2.10}$$

$$\delta L^2(\Omega) = \left\{\delta f : f \in L^2(\Omega)\right\}, \tag{2.11}$$

and we define the space of harmonic functions

$$\mathfrak{h} = \left\{\boldsymbol{v} \in H(\mathrm{div};\Omega) : \nabla\cdot\boldsymbol{v} = \delta^\perp\cdot\boldsymbol{v} = 0\right\}, \tag{2.12}$$

which is a finite-dimensional space with dimension depending only on the topology of $\Omega$. For example, for a simply-connected domain, the dimension is 2 (and the harmonic functions are constant vector fields).

The symbol $\oplus$ indicates that each of the components of $H(\mathrm{div};\Omega)$ are orthogonal in the $L^2$ inner product, i.e., for all $\boldsymbol{u} \in \nabla^\perp H^1(\Omega)$, $\boldsymbol{v} \in \mathfrak{h}$, $\boldsymbol{w} \in \delta L^2(\Omega)$,

$$\int_\Omega \boldsymbol{u}\cdot\boldsymbol{v}\,\mathrm{d}\boldsymbol{x} = \int_\Omega \boldsymbol{u}\cdot\boldsymbol{w}\,\mathrm{d}\boldsymbol{x} = \int_\Omega \boldsymbol{v}\cdot\boldsymbol{w}\,\mathrm{d}\boldsymbol{x} = 0. \tag{2.13}$$

Another important and related de Rham complex is defined on subspaces with boundary conditions. Functions in $H^1(\Omega)$ have enough regularity to define values on the boundary $\partial\Omega$ (interpreted as elements of $L^2(\partial\Omega)$. Thus we can define $\mathring{H}^1(\Omega)$ as

$$\mathring{H}^1(\Omega) = \left\{\psi \in H^1(\Omega) : \psi|_{\partial\Omega} = 0\right\}. \tag{2.14}$$

Similarly, vector fields $\boldsymbol{u} \in H(\mathrm{div};\Omega)$ have enough regularity to define normal components $\boldsymbol{u}\cdot\boldsymbol{n}$ on the boundary, where $\boldsymbol{n}$ is the unit outward pointing normal to $\partial\Omega$. Then we can define $\mathring{H}(\mathrm{div};\Omega)$ as

$$\mathring{H}(\mathrm{div};\Omega) = \{\boldsymbol{u} \in H(\mathrm{div};\Omega) : \boldsymbol{u}\cdot\boldsymbol{n}|_{\partial\Omega} = 0\}. \tag{2.15}$$

Functions in $L^2(\Omega)$ do not have enough regularity to define boundary values of any kind.

Similarly as before, we define corresponding dual operators

$$\delta : L^2(\Omega) \to \mathring{H}(\mathrm{div};\Omega), \tag{2.16}$$

$$\delta^\perp : \mathring{H}(\mathrm{div};\Omega) \to \mathring{H}^1(\Omega), \tag{2.17}$$

via

$$\int_\Omega \boldsymbol{w}\cdot\delta\phi\,\mathrm{d}\boldsymbol{x} = -\int_\Omega \nabla\cdot\boldsymbol{w}\phi\,\mathrm{d}\boldsymbol{x}, \quad \forall\boldsymbol{w} \in \mathring{H}(\mathrm{div};\Omega), \tag{2.18}$$

$$\int_\Omega \gamma\delta^\perp\cdot\boldsymbol{u}\,\mathrm{d}\boldsymbol{x} = -\int_\Omega \nabla^\perp\gamma\cdot\boldsymbol{u}\,\mathrm{d}\boldsymbol{x}, \quad \forall\gamma \in \mathring{H}^1(\Omega). \tag{2.19}$$

These dual operators with boundary conditions do not coincide with the $\nabla$ and $\nabla^\perp\cdot$ operators for smooth functions. We then have the de Rham complex

$$0 \to \mathring{H}^1(\Omega) \xrightarrow{\nabla^\perp} \mathring{H}(\mathrm{div};\Omega) \xrightarrow{\nabla\cdot} L^2(\Omega) \to 0, \tag{2.20}$$

and corresponding Helmholtz/Hodge decomposition,

$$\mathring{H}(\mathrm{div};\Omega) = \nabla^\perp\mathring{H}^1(\Omega) \oplus \mathring{\mathfrak{h}} \oplus \delta L^2(\Omega), \tag{2.21}$$

where

$$\mathring{\mathfrak{h}} = \left\{\boldsymbol{v} \in \mathring{H}(\mathrm{div};\Omega) : \nabla\cdot\boldsymbol{v} = \delta^\perp\cdot\boldsymbol{v} = 0\right\}, \tag{2.22}$$

are the harmonic functions that satisfy the boundary condition (this space has dimension zero for a simply connected domain).

Now we introduce compatible finite element spaces that mimic the above structure at the finite dimensional level. Let $V_0 \subset H^1$, $V_1 \subset H(\mathrm{div})$, and $V_2 \subset L^2$ be a sequence of global finite element spaces defined on a tessellation, $\mathcal{T}_h$, of $\Omega$. We say these spaces are *compatible* if:

1. for every $\psi \in V_0$, $\nabla^\perp \psi \in V_1$;
2. for every $\boldsymbol{u} \in V_1$, $\nabla \cdot \boldsymbol{u} \in V_2$; and
3. there exist bounded projections $\pi_i$, $i = 0, 1, 2$, such that the following diagram commutes.

$$\begin{array}{ccccc} H^1 & \xrightarrow{\nabla^\perp} & H(\mathrm{div}) & \xrightarrow{\nabla\cdot} & L^2 \\ \downarrow{\scriptstyle \pi_0} & & \downarrow{\scriptstyle \pi_1} & & \downarrow{\scriptstyle \pi_2} \\ V_0 & \xrightarrow{\nabla^\perp} & V_1 & \xrightarrow{\nabla\cdot} & V_2 \end{array} \tag{2.23}$$

The use of commutative diagrams like (2.23) to prove stability and convergence of mixed finite element methods for elliptic problems is well known, see Boffi et al. (2013).

As for all finite element spaces, the definition of these compatible spaces requires two pieces of information:

1. The polynomial space $P(K)$ that defines the finite element spaces restricted to a single cell $K$.
2. The continuity conditions between cells.

For a finite element space to be a subspace of $H^1(\Omega)$, it must be continuous between cells; this is just enough continuity to define the weak gradient operator. For a finite element space to be a subspace of $H(\mathrm{div};\Omega)$, only the edge-normal components $\boldsymbol{u} \cdot \boldsymbol{n}$ need to be continuous (where $\boldsymbol{n}$ is the unit normal vector to the edge between two cells); this is just enough continuity to define the weak divergence. There are no continuity requirements for $L^2(\Omega)$ finite element spaces.

In order to be implementable in practice, the polynomial space $P(K)$ needs to have a geometric decomposition that is consistent with the continuity conditions. A geometric decomposition is a basis for $P(K)$ that associates each basis function with a geometric entity of $K$ (i.e., a vertex, an edge, or $K$ itself). For $H^1(\Omega)$ functions, there are two requirements:

1. All basis functions not associated with a vertex must vanish at that vertex.
2. All basis functions not associated with an edge (or the vertices on the edge) must vanish on that edge.

Continuity can then be ensured by requiring that when two neighbouring cells share vertices or edges, the corresponding basis functions must have the same basis coefficients. For $H(\mathrm{div})$ functions, there is the requirement that all basis functions not associated with an edge (or the vertices on the edge) must have vanishing normal components on that edge. Similarly, continuity of normal components can then be ensured by requiring that basis functions on shared edges must have the same basis coefficients.

These continuity conditions also mean that it is possible to define subspaces with zero boundary conditions, i.e., $\mathring{V}_0(\Omega) \subset \mathring{H}^1(\Omega)$ and $\mathring{V}_1(\Omega) \subset \mathring{H}(\text{div};\Omega)$; the projection operators $\pi_1$ and $\pi_2$ also preserve the boundary conditions. Then the diagram (2.23) still holds with subspaces appropriately substituted.

The standard approach to describing the geometric basis for $P(K)$ is to specify a basis for the dual space $P'(K)$, the space of linear mappings from $P(K) \to \mathbb{R}$. Typically this basis consists of point evaluations (or evaluations of components for vector-valued spaces) or integrals along edges or over $K$. The elements of this basis are referred to as the *degrees of freedom* of $P(K)$. Diagrams illustrating $P'(K)$ for the $P_2, \text{BDM}_1, P_0^{(\text{DG})}$ compatible spaces are shown in Figure 2.1, with an illustration of how global continuity conditions are enforced in Figure 2.2.

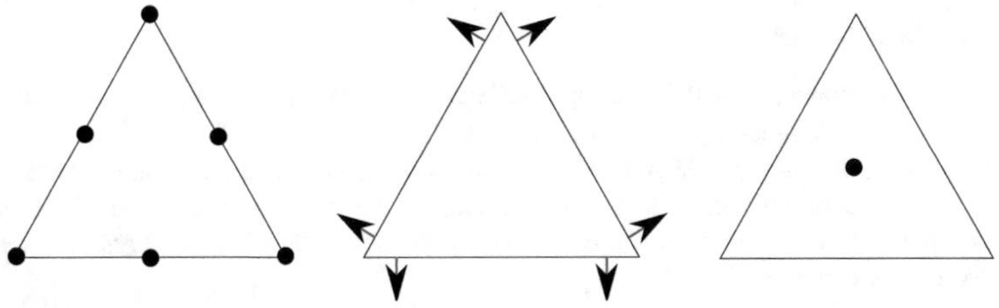

**Fig. 2.1** Diagram showing the degrees of freedom for the $P_2 \subset H^1(K)$ (quadratic functions), $\text{BDM}_1 \subset H(\text{div})(K)$ (linear vector fields), $P_0^{(\text{DG})} \subset L^2$ (constant functions) spaces on a triangle $K$. The dots indicate point evaluations whilst the arrows indicate edge-normal component evaluations. For $P_2$, we see that each edge has 3 point evaluations, enough to determine a quadratic function restricted to the edge, and hence no other point evaluations are required to determine the function value there. Hence the $P_2$ basis functions not associated to that edge will vanish there. Similarly, for $\text{BDM}_1$, each edge has 2 normal component evaluations, enough to determine the linear function corresponding to the normal component on that edge, and so all other basis functions will have vanishing normal components there. The gradient of a quadratic function gives a linear vector field, and the divergence of a linear vector field is constant, hence we can define a de Rham complex from these spaces

Some possible choices on triangular meshes for the $(V_0, V_1, V_2)$ tuple are:

- The family $(P_r, \text{BDM}_{r-1}, P_{r-2}^{(\text{DG})}), r \geq 2$, where $P_r$ are the continuous Lagrange elements of degree $r$, $\text{BDM}_{r-1}$ is the Brezzi-Douglas-Marini element of degree $r-1$, and $P_{r-2}^{(\text{DG})}$ are the discontinuous Lagrange elements of degree $r-2$.
- The family $(P_r, \text{RT}_{r-1}, P_{r-1}^{(\text{DG})}), r > 0$. Where, the space $RT_{r-1}$ denotes the Raviart-Thomas elements of degree $r-1$.

On quadrilateral meshes, we can choose:

- The family $(Q_r, \text{RTc}_{r-1}^{f}, Q_{r-1}^{(\text{DG})})$, where $Q_r$ denotes the continuous Lagrange elements on quadrilaterals of degree $\leq r$, $Q_{r-1}^{(\text{DG})}$ is the discontinuous Lagrange elements of degree $\leq r-1$, and $\text{RTc}_{r-1}^{f}$ denotes the Raviart-Thomas quadrilateral elements.

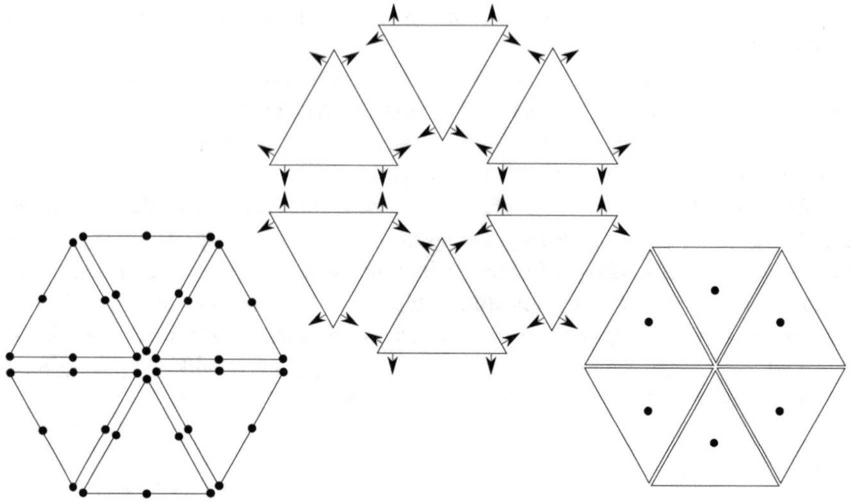

**Fig. 2.2** Diagram showing how to "glue together" degrees of freedom in neighbouring cells for the $P_2 \subset H^1(K)$ (quadratic functions), $\mathrm{BDM}_1 \subset H(\mathrm{div})(K)$ (linear vector fields), $P_0^{(\mathrm{DG})} \subset L^2$ (constant functions) spaces on a triangle $K$. For $P_2$, the values at vertices and edges are shared between neighbouring elements. This ensures global continuity of the finite element space. For $\mathrm{BDM}_1$ the normal components are shared between elements that share an edge. This ensures that the global (weak) divergence can be evaluated

For a complete description of these finite element spaces, see Arnold and Logg (2014).

One important thing to note is that whilst compatible finite element spaces allow for the operators $\nabla^\perp$ and $\nabla\cdot$ to map exactly between the finite element spaces $V_0$, $V_1$, $V_2$, the dual operators $\delta$ and $\delta^\perp$ are approximated by $\widehat{\delta}$ and $\widehat{\delta^\perp}$ obtained from definitions above with trial and test functions restricted to the finite element spaces appropriately.

Let $P_0$, $P_1$, and $P_2$ be $L^2$-projection operators into the compatible spaces $V_0$, $V_1$, and $V_2$ respectively, i.e.

$$\int_\Omega \gamma P_0(f)\,\mathrm{d}\boldsymbol{x} = \int_\Omega \gamma f\,\mathrm{d}\boldsymbol{x}, \text{ for all } \gamma \in V_0, \tag{2.24}$$

with $P_1$ and $P_2$ defined analogously. Then the weak operators $\widetilde{\delta^\perp}\cdot$ and $\widetilde{\delta}$ satisfy

$$\widetilde{\delta^\perp}\cdot(P_1(\boldsymbol{v})) = P_0\left(\delta^\perp\cdot\boldsymbol{v}\right), \tag{2.25}$$

$$\widetilde{\delta}(P_2(f)) = P_1(\delta f), \tag{2.26}$$

where $\boldsymbol{v}$ and $f$ are arbitrary functions; the dual differential operators commute with $L^2$-projections into the compatible function spaces.

## 2.2 Discontinuous Galerkin Scalar Advection

The advection of a scalar field $q(\boldsymbol{x},t)$ by a known velocity field $\boldsymbol{u}(\boldsymbol{x},t)$ is described by the equation

$$\frac{\partial q}{\partial t} + (\boldsymbol{u}\cdot\nabla)q = 0. \tag{2.27}$$

The field $q$ is known initially:

$$q(\boldsymbol{x},0) = q_0(\boldsymbol{x}). \tag{2.28}$$

If the domain has a boundary then $q$ must be known for all time on the subset of the boundary $\partial\Omega$ in which $\boldsymbol{u}$ is directed towards the interior of the domain:

$$q(\boldsymbol{x},t) = q_{\text{in}}(\boldsymbol{x},t) \quad \text{on } \partial\Omega_{\text{inflow}}. \tag{2.29}$$

The advection equation is used to describe the evolution of a passive tracer in some background flow. It is well known that a standard (continuous Galerkin) finite element approach to this equation results in oscillations and unphysical solutions. A common approach, therefore, is to represent the advected quantity $q$ in a discontinuous finite element space – one that has no continuity requirements at cell boundaries.

We show at the end of this section that the methods we describe lead to an almost-identical numerical scheme for the continuity equation, which governs the evolution of the height field in the shallow water equations, per (1.50), and the density field in the compressible Euler equations, per (1.2). While the potential temperature directly satisfies an advection equation in the Boussinesq and Euler equations, per (1.22) and (1.3), for numerical reasons this variable is not in a fully-discontinuous finite element space. A more specialised advection scheme for this variable is described in Section 2.4.4.

Let $V$ denote an appropriate space of discontinuous finite element functions, with $q \in V$. A weak form of the continuous equation in each cell $K$ is obtained by multiplying by a test function $\phi_K \in V_K$ and integrating over the cell:

$$\int_K \phi_K \frac{\partial q}{\partial t}\,\mathrm{d}\boldsymbol{x} + \int_K \phi_K (\boldsymbol{u}\cdot\nabla)q\,\mathrm{d}\boldsymbol{x} = 0, \quad \text{for all } \phi_K \in V_K, \tag{2.30}$$

where we explicitly introduce the subscript $K$ since the test functions $\phi_K$ are local to each cell. At this point, there is no coupling between neighbouring cells. Using integration by parts on the second term gives

$$\int_K \phi_K \frac{\partial q}{\partial t}\,\mathrm{d}\boldsymbol{x} = \int_K q\nabla\cdot(\phi_e \boldsymbol{u})\,\mathrm{d}\boldsymbol{x} - \int_{\partial K} \phi_K q \boldsymbol{u}\cdot\boldsymbol{n}_K\,\mathrm{d}s, \quad \text{for all } \phi_K \in V_K, \tag{2.31}$$

where $\boldsymbol{n}_K$ is an outward-pointing unit normal.

The value of $q$ is not well-defined on mesh facets, since $q$ is in a discontinuous finite element space. It is possible to choose sophisticated 'flux functions', but here we simply choose the upwind value of $q$ with respect to the velocity field, $\widetilde{q}$. This introduces coupling between neighbouring mesh cells. There are three types of facets that we may encounter:

1. Interior facets. Here, the value of $q$ from the upstream side, denoted $\widetilde{q}$, is used.
2. Inflow boundary facets, where $\boldsymbol{u}$ points towards the interior. Here, the upstream value is the prescribed boundary value $q_{\text{in}}$.
3. Outflow boundary facets, where $\boldsymbol{u}$ points towards the outside. Here, the upstream value is the interior solution value $q$.

We now sum our earlier expression (2.31) over all cells $K$. The cell integrals are easy to express, since $\sum_K \int_K \cdot \, \mathrm{d}\boldsymbol{x} = \int_\Omega \cdot \, \mathrm{d}\boldsymbol{x}$. The interior facet integrals are more subtle, since each facet in the set of interior facets, which we hereby denote $\Gamma$, appears twice in the sum $\sum_K \int_{\partial K}$. In other words, contributions arise from both of the neighbouring cells. The exterior facet integrals are easier to handle, since each facet in the set of exterior facets $\partial\Omega$ appears exactly once in the sum $\sum_K \int_{\partial K}$. The full equations are then

$$\boxed{\begin{aligned}\int_\Omega \phi \frac{\partial q}{\partial t} \, \mathrm{d}\boldsymbol{x} &= \int_\Omega q \nabla \cdot (\phi \boldsymbol{u}) \, \mathrm{d}\boldsymbol{x} - \int_\Gamma [\![ \phi \boldsymbol{u} ]\!] \widetilde{q} \, \mathrm{d}S \\ &\quad - \int_{\partial\Omega_{\text{inflow}}} \phi q_{\text{in}} \boldsymbol{u} \cdot \boldsymbol{n} \, \mathrm{d}s - \int_{\partial\Omega_{\text{outflow}}} \phi q \boldsymbol{u} \cdot \boldsymbol{n} \, \mathrm{d}s,\end{aligned}} \tag{2.32}$$

for all $\phi \in V$. We use $[\![ \cdot ]\!]$ to denote the "jump" of a vector quantity $\boldsymbol{v}$ across a facet:

$$[\![ \boldsymbol{v} ]\!] = \boldsymbol{v}^+ \cdot \boldsymbol{n}^+ + \boldsymbol{v}^- \cdot \boldsymbol{n}^-. \tag{2.33}$$

Here, + and – are arbitrary labels for the cells on each side of the facet, and $\boldsymbol{n}^\pm$ is the unit normal vector pointing outwards from the respective cell. The $\nabla$ in the second term of (2.32) is technically a 'broken' gradient operator, i.e., the usual gradient restricted to each cell.

A closely-related equation is the scalar continuity equation

$$\frac{\partial q}{\partial t} + \nabla \cdot (\boldsymbol{u} q) = 0. \tag{2.34}$$

Following the same steps as above, integration by parts in a single cell $K$ gives a slightly different cell integral $\int_K$ but an identical surface integral $\int_{\partial K}$. The resulting global weak form is

$$\boxed{\begin{aligned}\int_\Omega \phi \frac{\partial q}{\partial t} \, \mathrm{d}\boldsymbol{x} &= \int_\Omega q (\boldsymbol{u} \cdot \nabla) \phi \, \mathrm{d}\boldsymbol{x} - \int_\Gamma [\![ \phi \boldsymbol{u} ]\!] \widetilde{q} \, \mathrm{d}S \\ &\quad - \int_{\partial\Omega_{\text{inflow}}} \phi q_{\text{in}} \boldsymbol{u} \cdot \boldsymbol{n} \, \mathrm{d}s - \int_{\partial\Omega_{\text{outflow}}} \phi q \boldsymbol{u} \cdot \boldsymbol{n} \, \mathrm{d}s.\end{aligned}} \tag{2.35}$$

As a timestepping scheme, a suitable choice for either of these equations is the three-stage strong-stability-preserving Runge-Kutta (SSPRK) scheme of Shu and Osher (1988). For an equation in the abstract form

$$\frac{\partial q}{\partial t} = \mathcal{L}(q), \tag{2.36}$$

the scheme is

$$q^{(1)} = q^n + \Delta t \mathcal{L}(q^n) \tag{2.37}$$

$$q^{(2)} = \frac{3}{4} q^n + \frac{1}{4}(q^{(1)} + \Delta t \mathcal{L}(q^{(1)})) \tag{2.38}$$

$$q^{n+1} = \frac{1}{3} q^n + \frac{2}{3}(q^{(2)} + \Delta t \mathcal{L}(q^{(2)})). \tag{2.39}$$

This numerical scheme is suitable for arbitrary order discontinuous finite elements. However, if elements of polynomial order $p > 0$ are used and sharp interfaces are present, oscillations are likely to appear. If this is undesirable, a slope limiter can be used, such as that described in Kuzmin (2010). For the SSPRK properties to formally hold, the limiter should be applied at every stage of the scheme.

## 2.3 Two-Dimensional Shallow Water System

In this section, we construct a discretisation of the shallow water equations. We start with the linear shallow water equations, using a discretisation based on the principles of compatible finite elements. We prove that the discretisation inherits several desirable properties of the continuous equations. We then extend this to a discretisation of the nonlinear shallow water equations. Both the linear and nonlinear shallow water equations are combined with an implicit timestepping scheme.

### 2.3.1 Semi-discrete Formulation of the Linearised Shallow Water System

A semi-discretisation of the linear shallow water equations, based on compatible finite elements, was presented in Cotter and Shipton (2012). This discretisation has several desirable properties, which we summarise here. Most notably, it leads to steady geostrophic modes on arbitrary meshes. It also inherits a number of conservation properties from the continuous equations, and doesn't suffer from spurious pressure modes.

The linear shallow water equations in a rotating frame of reference were given in (1.57) and (1.58),

$$\frac{\partial \boldsymbol{u}}{\partial t} + f\boldsymbol{u}^{\perp} = -g\nabla h, \tag{2.40}$$

$$\frac{\partial h}{\partial t} + H\nabla \cdot \boldsymbol{u} = 0, \tag{2.41}$$

where $\boldsymbol{u}$ and $h$ are the linearised fluid velocity and depth, and $\boldsymbol{u}^{\perp} = \hat{\boldsymbol{z}} \times \boldsymbol{u}$. We take $\boldsymbol{u}$ to be in a finite element space $V_1 \subset H(\mathrm{div})$, and $h$ to be in a finite element space $V_2 \subset L^2$, assuming that $V_1$ and $V_2$ are compatible, as defined in Section 2.1. We then follow the usual finite element approach of multiplying the equations by arbitrary test functions $\boldsymbol{w} \in V_1$ and $\phi \in V_2$, integrating over the domain $\Omega$, and using integration by parts to move the derivative off the discontinuous variable $h$. If the domain has no boundary (e.g., a periodic rectangle, or the surface of a sphere), the resulting boundary integrals can be omitted. On such a domain, a discretisation of (2.40) and (2.41), in space only, is

$$\int_{\Omega} \boldsymbol{w} \cdot \frac{\partial \boldsymbol{u}}{\partial t} \, \mathrm{d}\boldsymbol{x} + \int_{\Omega} \boldsymbol{w} \cdot (f \boldsymbol{u}^{\perp}) \, \mathrm{d}\boldsymbol{x} = \int_{\Omega} (\nabla \cdot \boldsymbol{w}) g h \, \mathrm{d}\boldsymbol{x}, \tag{2.42}$$

$$\int_{\Omega} \phi \frac{\partial h}{\partial t} \, \mathrm{d}\boldsymbol{x} + H \int_{\Omega} \phi \nabla \cdot \boldsymbol{u} \, \mathrm{d}\boldsymbol{x} = 0, \tag{2.43}$$

for all $\boldsymbol{w} \in V_1$ and $\phi \in V_2$. We explore several properties of this discretisation in the following subsections.

**Conservation of energy**

In Section 1.1.4, we saw that the linear shallow water equations conserve energy, per 1.60. Our semi-discretisation inherits this property, as we may verify directly,

$$\begin{aligned}
\frac{\partial E}{\partial t} &= \int H \boldsymbol{u} \cdot \frac{\partial \boldsymbol{u}}{\partial t} + g h \frac{\partial h}{\partial t} \, \mathrm{d}\boldsymbol{x}, \\
&= \int H(\nabla \cdot \boldsymbol{u}) g h - \underbrace{H f \boldsymbol{u} \cdot \boldsymbol{u}^{\perp}}_{=0} - g h H \nabla \cdot \boldsymbol{u} \, \mathrm{d}\boldsymbol{x} = 0,
\end{aligned}$$

where we used $H\boldsymbol{u}$ and $gh$ as test functions in (2.42) and (2.43) respectively. This is possible when $H$ and $g$ are constants, but the proof can be extended to the variable $H$ and $g$ case (Cotter and Kirby, 2016). The perturbation energy $E$ is therefore conserved.

**Local mass conservation**

Local mass conservation is a desirable, and sometimes essential, property for geophysical models since it prevents spurious loss or gain of mass. In our compatible finite element formulation, the discontinuous $V_2$ space contains the "local indicator functions": functions which are 1 in a particular element, and 0 in all others. Taking $\phi$ to be the local indicator function for an element $K$, (2.43) gives

$$\frac{\mathrm{d}}{\mathrm{d}t} \int_K h \, \mathrm{d}\boldsymbol{x} + H \int_K \nabla \cdot \boldsymbol{u} \, \mathrm{d}\boldsymbol{x} = 0. \tag{2.44}$$

Applying Gauss's theorem and rearranging then gives

$$\frac{\mathrm{d}}{\mathrm{d}t}\int_K h\,\mathrm{d}\boldsymbol{x} = -H\int_{\partial K} \boldsymbol{u}\cdot\boldsymbol{n}\,\mathrm{d}S, \tag{2.45}$$

where $\boldsymbol{n}$ is the unit outward normal vector for the boundary $\partial K$. The 'mass' (total fluid volume) in each cell therefore only changes due to fluxes through the boundary of the cell. We now show that any pair of neighbouring cells have a consistent view of the mass flux between them. Since $V_1$ defines a space of vector-valued functions with continuous normal components on cell facets, neighbouring cells agree on the value of $\boldsymbol{u}\cdot\boldsymbol{n}$ on the boundary between them. The cells therefore agree on the flux of $h$ through the boundary, implying local mass conservation.

**Absence of spurious pressure modes**

A principal reason the Arakawa C-grid is used in finite difference methods for atmosphere and ocean flows is that the discretisation does not suffer from a corrupting pressure mode. Such a mode can rapidly pollute the numerical solution in the presence of nonlinearities, boundary conditions, and forcing. In our case, pressure modes are related to the discretised gradient $\widetilde{\nabla} h$, where $h \in V_2$, and is defined via integration by parts to satisfy

$$\int_\Omega \boldsymbol{w}\cdot\widetilde{\nabla} h\,\mathrm{d}\boldsymbol{x} = -\int_\Omega \nabla\cdot\boldsymbol{w}\,h\,\mathrm{d}\boldsymbol{x}, \quad \text{for all } \boldsymbol{w}\in V_1. \tag{2.46}$$

Intuitively, a spurious pressure mode is a field $h$ which is not uniform, but whose discrete gradient is apparently zero: $\widetilde{\nabla} h \approx 0$. More formally, we can say that the finite element pair $(V_1, V_2)$ is free of spurious pressure modes if there exists a constant $\gamma$, independent of the maximal diameter[1] of the tessellation, such that for all $h \in V_2$, there is a corresponding $\boldsymbol{u} \in V_1$ with

$$\int_\Omega h\nabla\cdot\boldsymbol{u}\,\mathrm{d}\boldsymbol{x} \geq \gamma\|h\|_{L^2}\|\nabla\cdot\boldsymbol{u}\|_{L^2}. \tag{2.47}$$

This is identical to the classical finite element *inf–sup* condition. Traditionally, this has been proven on an ad-hoc basis for individual pairs of finite element spaces $(V_1, V_2)$. However, where $V_1$ and $V_2$ are generated from the ideas of finite element exterior calculus (FEEC), the condition is automatically satisfied. The details of this result are discussed by Arnold (2006 [§7]).

[1] For triangles in a triangulation, the maximum edge length.

### Steady geostrophic modes

Another desirable property is that the system should support steady, geostrophically-balanced states. The proof that this holds for our scheme follows from the existence of a discrete Helmholtz decomposition of functions in $V_1$. We introduce the compatible finite element space $V_0 \subset H^1$. For any $\psi \in V_0$ and $\phi \in V_2$, we then have

$$\int_\Omega \nabla^\perp \psi \cdot \widetilde{\nabla} \phi \, \mathrm{d}\boldsymbol{x} = \int_\Omega \left(\nabla \cdot \nabla^\perp \psi\right) \phi \, \mathrm{d}\boldsymbol{x} = 0. \tag{2.48}$$

This uses various compatible finite element identities: $\psi \in V_0 \implies \nabla^\perp \psi \in V_1$. We can then apply (2.46), taking $\boldsymbol{w} = \nabla^\perp \psi$. Finally, the vector calculus identity $\nabla \cdot \nabla^\perp \equiv 0$ leads to (2.48), which shows that $\nabla^\perp : V_0 \to V_1$ and $\widetilde{\nabla} : V_2 \to V_1$ map onto orthogonal subspaces of $V_1$. This means there exists an injective mapping between elements of $V_1$ and elements of $V_0 \times V_2$, defining a discrete Helmholtz decomposition:

$$\boldsymbol{u} = \nabla^\perp \psi + \widetilde{\nabla} \phi + \boldsymbol{h}, \tag{2.49}$$

for $\psi \in V_0$, $\phi \in V_2$, and $\boldsymbol{h} \in \mathcal{H} \subset V_1$ is a harmonic velocity field, where

$$\mathcal{H} = \left\{ \boldsymbol{w} \in V_1 \mid \nabla \cdot \boldsymbol{w} = 0, \int_\Omega \boldsymbol{w} \cdot \nabla^\perp \psi \, \mathrm{d}\boldsymbol{x} = 0, \text{ for all } \psi \in V_0 \right\}. \tag{2.50}$$

Again, this construction is proven more generally using the results of FEEC in Arnold et al. (2006, 2010).

To show that the formulation (2.42)–(2.43) admits steady geostrophic modes, we need to show that any divergence-free velocity field $\boldsymbol{u}$ has a corresponding depth field $h$ such that the system is in equilibrium. Ignoring the harmonic component of the velocity field, which is just a constant, we have $\boldsymbol{u} = \nabla^\perp \psi$ for some $\psi \in V_0$. We then define $h$ to satisfy

$$g \int_\Omega \phi h \, \mathrm{d}\boldsymbol{x} = \int_\Omega f \phi \psi \, \mathrm{d}\boldsymbol{x}, \tag{2.51}$$

for all $\phi \in V_2$. With this choice of $h$, the velocity field is steady: the velocity evolution equation (2.42) gives, for all test functions $\boldsymbol{w} \in V_1$,

$$\begin{aligned}
\frac{\mathrm{d}}{\mathrm{d}t} \int_\Omega \boldsymbol{w} \cdot \boldsymbol{u} \, \mathrm{d}\boldsymbol{x} &= -\int_\Omega \boldsymbol{w} \cdot \left(f \boldsymbol{u}^\perp\right) \mathrm{d}\boldsymbol{x} + g \int_\Omega (\nabla \cdot \boldsymbol{w}) h \, \mathrm{d}\boldsymbol{x} \\
&= -\int_\Omega f \boldsymbol{w} \cdot \left(\nabla^\perp \psi\right)^\perp \mathrm{d}\boldsymbol{x} + g \int_\Omega (\nabla \cdot \boldsymbol{w}) h \, \mathrm{d}\boldsymbol{x} \\
&= \int_\Omega f \boldsymbol{w} \cdot \nabla \psi \, \mathrm{d}\boldsymbol{x} + g \int_\Omega (\nabla \cdot \boldsymbol{w}) h \, \mathrm{d}\boldsymbol{x} \\
&= -\int_\Omega f (\nabla \cdot \boldsymbol{w}) \psi \, \mathrm{d}\boldsymbol{x} + g \int_\Omega (\nabla \cdot \boldsymbol{w}) h \, \mathrm{d}\boldsymbol{x} \\
&= 0.
\end{aligned} \tag{2.52}$$

We use integration by parts between the third and fourth line; the functions have sufficient continuity. We also use the fact that $\nabla \cdot \boldsymbol{w} \in V_2$, and can hence take $\phi = \nabla \cdot \boldsymbol{w}$ in (2.51) to show that the two terms in the fourth line are equal and opposite. We can also show the depth field is steady: for all test functions $\phi \in V_2$,

$$\frac{\mathrm{d}}{\mathrm{d}t} \int_\Omega \phi h \,\mathrm{d}\boldsymbol{x} = -H \int_\Omega \phi \nabla \cdot \boldsymbol{u} \,\mathrm{d}\boldsymbol{x} = 0, \tag{2.53}$$

since $\boldsymbol{u}$ was assumed divergence-free. Therefore, this discretisation supports steady geostrophic modes.

### 2.3.2 Fully Discrete Formulation of the Linear Shallow Water Equations

On reasonably-uniform meshes, an implicit scheme is not necessary, and the semi-discretisation (2.40)–(2.41) can be efficiently combined with a simple explicit method. However, in preparation for the complicated implicit formulations used in three-dimensional models, we use implicit midpoint (trapezoidal) timestepping here. A further benefit is that this timestepping scheme preserves quadratic invariants, in this case the perturbation energy.

Given $\boldsymbol{u}^n$ and $h^n$ at time $t_n$, the fully discrete problem is to find $\boldsymbol{u}^{n+1} \in V_1$ and $h^{n+1} \in V_2$ at time $t_{n+1} = t_n + \Delta t$. These satisfy

$$\int_\Omega \boldsymbol{w} \cdot \boldsymbol{u}^{n+1} \,\mathrm{d}\boldsymbol{x} + \frac{\Delta t}{2} \int_\Omega \boldsymbol{w} \cdot f \left(\boldsymbol{u}^{n+1}\right)^\perp \,\mathrm{d}\boldsymbol{x} - \frac{g\Delta t}{2} \int_\Omega h^{n+1} \nabla \cdot \boldsymbol{w} \,\mathrm{d}\boldsymbol{x} = -R_{\boldsymbol{u}}^n[\boldsymbol{w}], \tag{2.54}$$

$$\frac{H\Delta t}{2} \int_\Omega \phi \nabla \cdot \boldsymbol{u}^{n+1} \,\mathrm{d}\boldsymbol{x} + \int_\Omega \phi h^{n+1} \,\mathrm{d}\boldsymbol{x} = -R_h^n[\phi], \tag{2.55}$$

for all $\boldsymbol{w} \in V_1$ and $\phi \in V_2$, where the right-hand side functionals are defined as

$$R_{\boldsymbol{u}}^n[\boldsymbol{w}] = \int_\Omega \boldsymbol{w} \cdot \boldsymbol{u}^n \,\mathrm{d}\boldsymbol{x} - \frac{\Delta t}{2} \int_\Omega \boldsymbol{w} \cdot f (\boldsymbol{u}^n)^\perp \,\mathrm{d}\boldsymbol{x} + \frac{g\Delta t}{2} \int_\Omega h^n \nabla \cdot \boldsymbol{w} \,\mathrm{d}\boldsymbol{x}, \tag{2.56}$$

$$R_h^n[\phi] = -\frac{H\Delta t}{2} \int_\Omega \phi \nabla \cdot \boldsymbol{u}^n \,\mathrm{d}\boldsymbol{x} + \int_\Omega \phi h^n \,\mathrm{d}\boldsymbol{x}. \tag{2.57}$$

The implementation of this scheme is discussed in Section 4.1.

### 2.3.3 Semi-discrete Formulation of the Nonlinear Shallow Water Equations

The nonlinear shallow water equations in a rotating frame were presented in (1.49)–(1.50). We use the form of the momentum equation given in (1.51):

$$\frac{\partial \boldsymbol{u}}{\partial t} + (\nabla^{\perp} \cdot \boldsymbol{u} + f)\boldsymbol{u}^{\perp} + \nabla\left(g(h+\eta_b) + \frac{1}{2}|\boldsymbol{u}|^2\right) = 0, \tag{2.58}$$

$$\frac{\partial h}{\partial t} + \nabla \cdot (\boldsymbol{u}h) = 0. \tag{2.59}$$

The variables $\boldsymbol{u}$ and $h$ are the velocity and fluid depth, and we have included a bathymetry term $\eta_b$. Our discretisation extends the discretisation of the linear shallow water equations in Section 2.3.1, and is related to the discretisation of the incompressible Euler equations given in Natale and Cotter (2018).

We use a discontinuous Galerkin method for the continuity equation (2.59), as was presented in Section 2.2. In the momentum equation (2.58), the main difficulty is in representing the $(\nabla^{\perp} \cdot \boldsymbol{u})\boldsymbol{u}^{\perp}$ term, since there is insufficient continuity in $\boldsymbol{u}$ to meaningfully evaluate the derivative. Instead, we take the inner product with a test function $\boldsymbol{w}$ and integrate by parts, choosing the upwind value of $\boldsymbol{u}$ on the cell boundaries:

$$-\int_{\Omega} \boldsymbol{u} \cdot \nabla^{\perp}(\boldsymbol{u}^{\perp} \cdot \boldsymbol{w})\,\mathrm{d}\boldsymbol{x} + \int_{\Gamma} \left[\boldsymbol{n}^{\perp}(\boldsymbol{u}^{\perp} \cdot \boldsymbol{w})\right] \cdot \widetilde{\boldsymbol{u}}\,\mathrm{d}S. \tag{2.60}$$

As in Section 2.2, $\Gamma$ denotes the set of interior facets in the mesh and $\widetilde{\boldsymbol{u}}$ denotes the upwind value of $\boldsymbol{u}$ on each facet. The $\nabla^{\perp}$ is again a (rotated) broken gradient. $[\boldsymbol{v}]$ denotes the quantity $\boldsymbol{v}^{+} + \boldsymbol{v}^{-}$.

Our semi-discrete finite element formulation is as follows: find $\boldsymbol{u} \in V_1$, $h \in V_2$ such that

$$\boxed{\begin{aligned} &\int_{\Omega} \boldsymbol{w} \cdot \frac{\partial \boldsymbol{u}}{\partial t}\,\mathrm{d}\boldsymbol{x} - \int_{\Omega} \boldsymbol{u} \cdot \nabla^{\perp}(\boldsymbol{u}^{\perp} \cdot \boldsymbol{w})\,\mathrm{d}\boldsymbol{x} + \int_{\Gamma} \left[\boldsymbol{n}^{\perp}(\boldsymbol{u}^{\perp} \cdot \boldsymbol{w})\right] \cdot \widetilde{\boldsymbol{u}}\,\mathrm{d}S \\ &\quad + \int_{\Omega} \boldsymbol{w} \cdot (f\boldsymbol{u}^{\perp})\,\mathrm{d}\boldsymbol{x} - \int_{\Omega} \nabla \cdot \boldsymbol{w} \left(g(h+\eta_b) + \frac{1}{2}|\boldsymbol{u}|^2\right)\mathrm{d}\boldsymbol{x} = 0, \quad (2.61) \\ &\int_{\Omega} \phi \frac{\partial h}{\partial t}\,\mathrm{d}\boldsymbol{x} - \int_{\Omega} h(\boldsymbol{u} \cdot \nabla)\phi\,\mathrm{d}\boldsymbol{x} + \int_{\Gamma} [\![\boldsymbol{u}\phi]\!]\widetilde{h}\,\mathrm{d}\boldsymbol{x} = 0, \quad (2.62) \end{aligned}}$$

for all test functions $\boldsymbol{w} \in V_1$, $\phi \in V_2$. This scheme has no particular conservation properties beyond local and global mass conservation.

For the interested reader, an energy-conserving formulation can be obtained by introducing a mass flux variable $\boldsymbol{F} \approx \boldsymbol{u}h$, defined as the $L^2$ projection $\boldsymbol{F} = \pi_1(\boldsymbol{u}h)$. One then replaces $\boldsymbol{u}$ by $\boldsymbol{F}/h$ in the momentum equation (2.61), and uses the continuity equation $\frac{\partial h}{\partial t} + \nabla \cdot \boldsymbol{F} = 0$. A scheme that conserves mass, energy, and potential enstrophy was given in McRae and Cotter (2014); this introduced extra variables representing mass flux and potential vorticity. Other sophisticated discretisations can be found in Bauer and Cotter (2018), Shipton et al. (2018).

### 2.3.4 Fully Discrete Formulation of the Nonlinear Shallow Water Equations

We discretise (2.61) and (2.62) using the implicit midpoint rule. This implies the residuals $R_{\boldsymbol{u}}[\boldsymbol{w};\boldsymbol{u}^{n+1},h^{n+1}]$ and $R_h[\phi;\boldsymbol{u}^{n+1},h^{n+1}]$ should equal zero for all $\boldsymbol{w} \in V_1$ and all $\phi \in V_2$, where

$$\begin{aligned} R_{\boldsymbol{u}}[\boldsymbol{w};\boldsymbol{u}^{n+1},h^{n+1}] = & \int_\Omega \boldsymbol{w} \cdot \frac{\boldsymbol{u}^{n+1} - \boldsymbol{u}^n}{\Delta t} \, \mathrm{d}\boldsymbol{x} \\ & - \int_\Omega \boldsymbol{u}^{n+\frac{1}{2}} \cdot \nabla^\perp \left( (\boldsymbol{u}^\perp)^{n+\frac{1}{2}} \cdot \boldsymbol{w} \right) \mathrm{d}\boldsymbol{x} \\ & + \int_\Gamma \left[ \boldsymbol{n}^\perp \left( (\boldsymbol{u}^\perp)^{n+\frac{1}{2}} \cdot \boldsymbol{w} \right) \right] \cdot \widetilde{\boldsymbol{u}}^{n+\frac{1}{2}} \, \mathrm{d}S \\ & + \int_\Omega \boldsymbol{w} \cdot f (\boldsymbol{u}^\perp)^{n+\frac{1}{2}} \, \mathrm{d}\boldsymbol{x} \\ & - \int_\Omega \nabla \cdot \boldsymbol{w} \left( g \left( h^{n+\frac{1}{2}} + \eta_b \right) + \frac{1}{2} |\boldsymbol{u}^{n+\frac{1}{2}}|^2 \right) \mathrm{d}\boldsymbol{x}, \end{aligned} \tag{2.63}$$

and

$$\begin{aligned} R_h[\phi;\boldsymbol{u}^{n+1},h^{n+1}] = & \int_\Omega \phi \, \frac{h^{n+1} - h^n}{\Delta t} \, \mathrm{d}\boldsymbol{x} \\ & - \int_\Omega h^{n+\frac{1}{2}} \left( \boldsymbol{u}^{n+\frac{1}{2}} \cdot \nabla \right) \phi \, \mathrm{d}\boldsymbol{x} \\ & + \int_\Gamma \left[\!\left[ \phi \boldsymbol{u}^{n+\frac{1}{2}} \right]\!\right] \widetilde{h}^{n+\frac{1}{2}} \, \mathrm{d}S, \end{aligned} \tag{2.64}$$

with $\boldsymbol{u}^{n+\frac{1}{2}} = \frac{1}{2}(\boldsymbol{u}^{n+1} + \boldsymbol{u}^n)$, and $h^{n+\frac{1}{2}} = \frac{1}{2}(h^{n+1} + h^n)$.

This discretisation is unconditionally stable, but requires the solution of a nonlinear coupled system of equations for $\boldsymbol{u}^{n+1}$ and $h^{n+1}$. One could use Newton's method directly, but this has some disadvantages. The Jacobian must be reassembled each iteration. If using a Krylov method to solve the linear system, the operator is difficult to precondition; if using a direct method, the factorisation must be recalculated repeatedly. Instead, we use an approximate Jacobian, obtained by linearising around a fixed background profile. This is state-independent, easier to precondition, and allows reuse of a sparse LU factorisation. To control the reduction of the nonlinear residual, we apply a number of Picard iterations towards the solution of the implicit midpoint system. As long as this approximate Jacobian is not too far from the true Jacobian, the nonlinear residuals $R_{\boldsymbol{u}}$ and $R_h$ can still be made arbitrarily small.

The quasi-Newton/Picard iteration scheme is summarised as follows. First, we define a sequence of approximations $\boldsymbol{v}^0, \boldsymbol{v}^1, \boldsymbol{v}^2, \ldots$, and $p^0, p^1, p^2, \ldots$ to $\boldsymbol{u}^{n+1}$ and $h^{n+1}$ respectively, with $\boldsymbol{v}^0 = \boldsymbol{u}^n$ and $p^0 = h^n$. Next, we define the increments $\delta \boldsymbol{v}^k$ and $\delta p^k$, with

$$\boldsymbol{v}^{k+1} = \boldsymbol{v}^k + \delta \boldsymbol{v}^k, \quad p^{k+1} = p^k + \delta p^k. \tag{2.65}$$

These increments are chosen to satisfy the linear finite element problem

$$\int_\Omega \boldsymbol{w} \cdot \delta \boldsymbol{v}^k \, \mathrm{d}\boldsymbol{x} + \frac{\Delta t}{2} \int_\Omega \boldsymbol{w} \cdot f \left(\delta \boldsymbol{v}^k\right)^\perp \mathrm{d}\boldsymbol{x} - \frac{g\Delta t}{2} \int_\Omega \delta p^k \nabla \cdot \boldsymbol{w} \, \mathrm{d}\boldsymbol{x} = -R_{\boldsymbol{u}}[\boldsymbol{w}; \boldsymbol{v}^k, p^k], \tag{2.66}$$

$$\frac{H\Delta t}{2} \int_\Omega \phi \nabla \cdot \delta \boldsymbol{v}^k \, \mathrm{d}\boldsymbol{x} + \int_\Omega \phi \delta p^k = -R_h[\phi; \boldsymbol{v}^k, p^k], \tag{2.67}$$

where the residual functionals $R_{\boldsymbol{u}}$ and $R_h$ are as defined above. Here, we have reused the Jacobian from the linear shallow water equations, per (2.54) and (2.55), equivalent to linearising about a base layer depth $H$ at rest. We implement this scheme in Section 4.3, making some further extensions.

## 2.4 Three-Dimensional Formulations

The two-dimensional formulations of Section 2.3 provides a useful stepping stone for the development of compatible finite element methods for geophysical flows. As in the previous section, we begin by defining a compatible sequence of finite element spaces in three dimensions. We note here that the extension of vertical grid staggering of finite difference methods to finite element methods, including higher-order elements, was introduced by Guerra and Ullrich (2016). For our purposes, we consider the use of three-dimensional spaces that have the same structure in the vertical.

### 2.4.1 Compatible Spaces in Three Dimensions

As before, a sequence of finite element spaces can be described as "compatible" if there exist bounded projections $\pi_i$, such that we have a commutative diagram (2.68), as described in Arnold (2006).

$$\begin{array}{ccccccc}
H^1 & \xrightarrow{\nabla} & H(\mathrm{curl}) & \xrightarrow{\nabla\times} & H(\mathrm{div}) & \xrightarrow{\nabla\cdot} & L^2 \\
\downarrow \pi_0 & & \downarrow \pi_1 & & \downarrow \pi_2 & & \downarrow \pi_3 \\
W_0 & \xrightarrow{\nabla} & W_1 & \xrightarrow{\nabla\times} & W_2 & \xrightarrow{\nabla\cdot} & W_3
\end{array} \tag{2.68}$$

We now formally define what we mean by "compatible" finite element spaces in three dimensions. Let $W_0 \subset H^1$, $W_1 \subset H(\text{curl})$, $W_2 \subset H(\text{div})$, and $W_3 \subset L^2$ be a sequence of finite element spaces. We say these spaces are *compatible* if the following properties hold:

1. for every $\psi \in W_0$, $\nabla\psi \in W_1$;
2. for every $\boldsymbol{q} \in W_1$, $\nabla \times \boldsymbol{q} \in W_2$;
3. for every $\boldsymbol{u} \in W_2$, $\nabla \cdot \boldsymbol{u} \in W_3$; and
4. there exist bounded projections $\pi_i$, $i = 0, 1, 2, 3$, such that the diagram (2.68) commutes.

In a compatible finite element discretisation of three-dimensional compressible systems (such as the Euler equations), velocities are chosen from the space $W_2$ and density from $W_3$. We will discuss how temperature is handled momentarily. In large scale models, the computational domain is quite thin, therefore cells have a much greater extent in the horizontal direction than the vertical. As a result, cells are aligned in vertical columns, otherwise any large horizontal errors in pressure would be coupled into the hydrostatic relation

$$\frac{\partial p}{\partial z} = -g\rho. \tag{2.69}$$

This causes the discretisation to generate spurious modes that rapidly degrade the state of hydrostatic balance. It was shown in Natale et al. (2016) that a vertically aligned mesh in tandem with compatible finite element spaces leads to a well-defined hydrostatic pressure.

### 2.4.2 Linearised Gravity Wave System

The following linear gravity wave problem for pressure $p$, velocity $\boldsymbol{u}$, and buoyancy $b$ can be obtained by linearising the the Boussinesq equations. For simplicity, we neglect the effects of rotation, but they can easily be incorporated into the resulting system. The model is given by

$$\frac{\partial \boldsymbol{u}}{\partial t} = -\nabla p + b\hat{z}, \tag{2.70}$$

$$\frac{\partial p}{\partial t} = -c^2 \nabla \cdot \boldsymbol{u}, \tag{2.71}$$

$$\frac{\partial b}{\partial t} = -N^2 \boldsymbol{u} \cdot \hat{z}, \tag{2.72}$$

where $c$ is the speed of sound and $N$ is the Brunt-Vaisälä frequency with $N^2 \equiv g\partial_z\theta_0/\theta_0$, and $\theta_0$ is a background profile for potential temperature. We enforce the strong boundary condition $\boldsymbol{u} \cdot \boldsymbol{n} = 0$ at the upper and lower boundary of the domain $\Omega$. We assume that $\Omega$ has no horizontal boundaries, such as a spherical shell.

To construct the relevant compatible finite element spaces, we consider the following one- and two-dimensional de Rham complexes

$$U_0 \xrightarrow{\frac{\mathrm{d}}{\mathrm{d}z}} U_1, \quad V_0 \xrightarrow{\nabla} V_1 \xrightarrow{\nabla \cdot} V_2. \tag{2.73}$$

Here, we are taking the one-dimensional complex to be a complex defined in the vertical direction. By taking the tensor product of the two-dimensional complex with the vertical complex, we can construct a three-dimensional complex

$$W_0 \xrightarrow{\nabla} W_1 \xrightarrow{\nabla\times} W_2 \xrightarrow{\nabla \cdot} W_3, \tag{2.74}$$

where

$$W_0 = V_0 \otimes U_0, \tag{2.75}$$

$$W_1 = [V_1 \otimes U_0] \oplus [V_0 \otimes U_1], \tag{2.76}$$

$$W_2 = [V_2 \otimes U_0] \oplus [V_1 \otimes U_1], \tag{2.77}$$

$$W_3 = V_2 \otimes U_1. \tag{2.78}$$

For velocity and pressure, we seek $\boldsymbol{u} \in \mathring{W}_2$, and $p \in W_3$. The space $\mathring{W}_2$ is the subspace of $W_2$ whose normal components vanish on the boundary of the domain – this enforces the strong boundary condition on $\boldsymbol{u}$. For the buoyancy $b$, we use the finite element space $W_b$, where $W_b$ is the scalar "vertical" part of $W_2$. Using the definition of $W_2$, $W_b$ is the product element $V_2 \otimes U_0$. Functions in $W_b$ are, in general, discontinuous in the horizontal direction, but continuous in the vertical. This combination of finite element spaces for $\boldsymbol{u}$ and $b$ is a finite element extension of the vertically staggered Charney–Phillips grid (Charney and Phillips, 1953).

The semi-discrete compatible finite element formulation of the gravity wave problem is then: find $\boldsymbol{u} \in \mathring{W}_2$, $p \in W_3$ and $b \in W_b$ such that

$$\int_\Omega \boldsymbol{w} \cdot \frac{\partial \boldsymbol{u}}{\partial t}\,\mathrm{d}\boldsymbol{x} - \int_\Omega \nabla \cdot \boldsymbol{w} p\,\mathrm{d}\boldsymbol{x} - \int_\Omega \boldsymbol{w} \cdot b\hat{\boldsymbol{z}}\,\mathrm{d}\boldsymbol{x} = 0, \tag{2.79}$$

$$\int_\Omega \phi \frac{\partial p}{\partial t}\,\mathrm{d}\boldsymbol{x} + c^2 \int_\Omega \phi \nabla \cdot \boldsymbol{u}\,\mathrm{d}\boldsymbol{x} = 0, \tag{2.80}$$

$$\int_\Omega \gamma \frac{\partial b}{\partial t}\,\mathrm{d}\boldsymbol{x} + N^2 \int_\Omega \gamma \boldsymbol{u} \cdot \hat{\boldsymbol{z}}\,\mathrm{d}\boldsymbol{x} = 0, \tag{2.81}$$

for all $\boldsymbol{w} \in \mathring{W}_2$, $\phi \in W_3$, and $\gamma \in W_b$. Note that all surface integrals arising from integrating by parts vanish due to the presence of the strong boundary conditions.

### 2.4.3 Implicit Discretisation of the Gravity Wave System

As with the two-dimensional nonlinear shallow water system of the previous section, we solve the problem by finding increments at time level $n$: $\delta \boldsymbol{u}^{(n)}$, $\delta p^{(n)}$ and $\delta b^{(n)}$, where

$$\boldsymbol{u}^{(n+1)} = \boldsymbol{u}^{(n)} + \delta \boldsymbol{u}^{(n)}, \tag{2.82}$$

$$p^{(n+1)} = p^{(n)} + \delta p^{(n)}, \tag{2.83}$$

$$b^{(n+1)} = b^{(n)} + \delta b^{(n)}. \tag{2.84}$$

In a similar manner, we use the $\theta$-method to define the intermediate quantities $\boldsymbol{u}^*$, $p^*$, and $b^*$ as the following:

$$\boldsymbol{u}^* = \boldsymbol{u}^{(n)} + (1-\theta)\delta \boldsymbol{u}^{(n)}, \tag{2.85}$$

$$p^* = p^{(n)} + (1-\theta)\delta p^{(n)}, \tag{2.86}$$

$$b^* = b^{(n)} + (1-\theta)\delta b^{(n)}. \tag{2.87}$$

For this example, we take $\theta = 1/2$. After substituting our definitions for the $\delta$-quantities, the intermediate quantities, and rearranging, the fully discrete system then reads: find the next increments $\delta \boldsymbol{u}^{(n+1)} \in \mathring{W}_2$, $\delta p^{(n+1)} \in W_3$, $\delta b^{(n+1)} \in W_b$ such that

$$\boxed{\begin{aligned}
\int_\Omega \boldsymbol{w} \cdot \delta \boldsymbol{u}^{(n+1)} \,\mathrm{d}\boldsymbol{x} - \frac{\Delta t}{2} \int_\Omega \nabla \cdot \boldsymbol{w} \delta p^{(n+1)} \,\mathrm{d}\boldsymbol{x} & \\
- \frac{\Delta t}{2} \int_\Omega \boldsymbol{w} \cdot \delta b^{(n+1)} \hat{\boldsymbol{z}} \,\mathrm{d}\boldsymbol{x} &= \mathcal{R}_{\boldsymbol{w}}, && (2.88)\\
\int_\Omega \phi \delta p^{(n+1)} + \frac{\Delta t}{2} c^2 \int_\Omega \phi \nabla \cdot \delta \boldsymbol{u}^{(n+1)} \,\mathrm{d}\boldsymbol{x} &= \mathcal{R}_\phi, && (2.89)\\
\int_\Omega \gamma \delta b^{(n+1)} \,\mathrm{d}\boldsymbol{x} + \frac{\Delta t}{2} N^2 \int_\Omega \gamma \delta \boldsymbol{u}^{(n+1)} \cdot \hat{\boldsymbol{z}} \,\mathrm{d}\boldsymbol{x} &= \mathcal{R}_\gamma, && (2.90)
\end{aligned}}$$

for all $\boldsymbol{w} \in \mathring{W}_2$, $\phi \in W_3$, $\gamma \in W_b$, where the residuals are defined as

$$\boxed{\begin{aligned}
\mathcal{R}_{\boldsymbol{w}} &:= \Delta t \int_\Omega \nabla \cdot \boldsymbol{w} p^{(n)} \,\mathrm{d}\boldsymbol{x} + \Delta t \int_\Omega \boldsymbol{w} \cdot b^{(n)} \hat{\boldsymbol{z}} \,\mathrm{d}\boldsymbol{x}, && (2.91)\\
\mathcal{R}_\phi &:= -\Delta t c^2 \int_\Omega \phi \nabla \cdot \boldsymbol{u}^{(n)} \,\mathrm{d}\boldsymbol{x}, && (2.92)\\
\mathcal{R}_\gamma &:= -\Delta t N^2 \int_\Omega \gamma \boldsymbol{u}^{(n)} \cdot \hat{\boldsymbol{z}} \,\mathrm{d}\boldsymbol{x}. && (2.93)
\end{aligned}}$$

Note that setting $\theta = 1/2$ corresponds to choosing implicit mid-point as the time discretisation. Since our problem is linear, this is also a special case of the Crank–Nicolson method (Crank and Nicolson, 1947).

### 2.4.4 An Advection Scheme for Potential Temperature

In the fully compressible system, as well as the nonlinear Boussinesq equations for the ocean, the advection of potential temperature appears in the equations as

$$\frac{D\theta}{Dt} = \frac{\partial \theta}{\partial t} + (\boldsymbol{u} \cdot \nabla)\theta = 0. \tag{2.94}$$

The temperature variable, much like the buoyancy term of the linearised gravity wave problem, is in a finite element space with only a partial degree of continuity; it is continuous between cells in the vertical, but discontinuous in the horizontal. We denote this space as $W_\theta$, and it is constructed similarly to $W_b$ of the previous section.

For $\theta \in W_\theta$, applying integration by parts in a column of cells $C$ gives

$$\int_C \gamma \frac{\partial \theta}{\partial t} \,\mathrm{d}\boldsymbol{x} - \int_C \theta \nabla \cdot (\gamma \boldsymbol{u}) \,\mathrm{d}\boldsymbol{x} + \int_{\partial C_z} \gamma \theta \boldsymbol{u} \cdot \boldsymbol{n} \,\mathrm{d}S = 0, \tag{2.95}$$

for all $\gamma \in W_\theta$, where $\partial C_z$ is the set of vertical facets of the cell column; integrals over the horizontal facets vanish due to the continuity of $W_\theta$. Taking the upwind value of $\theta$ and integrating over the entire mesh gives

$$\int_\Omega \gamma \frac{\partial \theta}{\partial t} \,\mathrm{d}\boldsymbol{x} - \int_\Omega \theta \nabla \cdot (\gamma \boldsymbol{u}) \,\mathrm{d}\boldsymbol{x} + \int_{\mathcal{E}_z^\circ} [\![ \gamma \boldsymbol{u} ]\!] \widetilde{\theta} \,\mathrm{d}S = 0, \tag{2.96}$$

for all $\gamma \in W_\theta$, where $\mathcal{E}_z^\circ$ is the set of vertical interior facets of the mesh. Now integrate (2.96) by parts again, this time without taking the upwind value of $\theta$, to obtain our advection equation

$$\int_\Omega \gamma \frac{\partial \theta}{\partial t} \,\mathrm{d}\boldsymbol{x} + \int_\Omega \gamma \boldsymbol{u} \cdot \nabla \theta \,\mathrm{d}\boldsymbol{x} + \int_{\mathcal{E}_z^\circ} [\![ \gamma \boldsymbol{u} ]\!] \widetilde{\theta} - [\![ \gamma \boldsymbol{u} \theta ]\!] \,\mathrm{d}S = 0, \tag{2.97}$$

for all $\gamma \in W_\theta$. Note that due to the continuity between vertical cells, the method is unstable and requires a stabilisation term. The Streamline Upwind Petrov-Galerkin (SUPG) method can be applied here, with vertically upwinded test functions,

$$\gamma \mapsto \gamma + \alpha \Delta z (\hat{\boldsymbol{w}} \cdot \nabla)\gamma, \tag{2.98}$$

where $\alpha$ is the SUPG parameter, $\Delta z$ is the vertical grid spacing, and $\hat{\boldsymbol{w}}$ is a unit vector field pointing in the direction of the vertical component of the velocity $\boldsymbol{u}$.

An alternative approach, which avoids using the SUPG scheme, is to embed $\theta$ into a fully discontinuous space, like $W_3$, and apply the standard DG advection scheme (2.32) using SSPRK; this approach is stable for suitably small time-steps. Finally, the resulting field is projected (via a standard Galerkin projection) back into $W_\theta$. We note that while slope limiters are not typically necessary for the temperature field, certain tracers may be stored in the space $W_\theta$ for physics-dynamics coupling. These tracers may indeed require the use of a slope limiter to avoid corrupted solutions.

Recent work on slope limiters for these partially-continuous spaces can be found in Cotter and Kuzmin (2016).

### 2.4.5 Nonlinear Compressible Equations

In this section, we apply the compatible finite element techniques discuss thus far on the fully compressible system for the atmosphere (summarised in Section 1.1.1). Compared with those equations, we make a few adjustments following Wood et al. (2014). We replace the geopotential term $\nabla\Phi$ by simply $g\hat{z}$, and we eliminate the pressure $p$ by replacing $\frac{1}{\rho}\nabla p$ with $c_p\theta\nabla\Pi$, where $\Pi$ is the Exner pressure

$$\Pi = \left(\frac{R_d\rho\theta}{p_0}\right)^{\frac{\kappa}{1-\kappa}}, \tag{2.99}$$

where $p_0$ is a reference pressure and $\kappa = R_d/c_p$. We also neglect any viscous forcing and source/sink terms. Our equations are then

$$\frac{\partial \boldsymbol{u}}{\partial t} + (\boldsymbol{u}\cdot\nabla)\boldsymbol{u} + 2\boldsymbol{\Omega}\times\boldsymbol{u} = -\theta\nabla\Pi - g\hat{z}, \tag{2.100}$$

$$\frac{\partial \theta}{\partial t} + \boldsymbol{u}\cdot\nabla\theta = 0, \tag{2.101}$$

$$\frac{\partial \rho}{\partial t} + \nabla\cdot(\rho\boldsymbol{u}) = 0. \tag{2.102}$$

Note that within the given formulation, the pressure can be determined via the equation of state $p = \rho R_d T$ combined with the definition of $\Pi$ in (2.99) and the definition of potential temperature:

$$\theta = T\left(\frac{p_0}{p}\right)^{\kappa}. \tag{2.103}$$

Due to the presence of the horizontally-discontinuous variable $\theta$, we obtain a non-vanishing surface term when integrating the momentum equation (2.100) by parts. Multiplying the pressure gradient term $c_p\theta\nabla\Pi$ by a test function and integrating by parts over a vertical column of cells $C$, we have

$$\int_C \boldsymbol{w}\cdot\left(c_p\theta\nabla\Pi\right)\,\mathrm{d}\boldsymbol{x} = \int_{\partial C_z} c_p\Pi\theta\boldsymbol{w}\cdot\boldsymbol{n}\,\mathrm{d}S - \int_C c_p\Pi\nabla_h\cdot(\theta\boldsymbol{w})\,\mathrm{d}\boldsymbol{x}, \tag{2.104}$$

where $\nabla_h$ is the broken gradient operator obtained by evaluating the gradient pointwise in each cell. Taking the column-local contributions (2.104) and summing over all columns of the mesh, we obtain

$$\int_\Omega \boldsymbol{w}\cdot\left(c_p\theta\nabla\Pi\right)\,\mathrm{d}\boldsymbol{x} = \int_{\mathcal{E}_z^\circ} c_p[\![\theta\boldsymbol{w}]\!]\{\!\{\Pi\}\!\}\,\mathrm{d}S - \int_\Omega c_p\Pi\nabla_h\cdot(\theta\boldsymbol{w})\,\mathrm{d}\boldsymbol{x}, \tag{2.105}$$

where $\{\!\{\cdot\}\!\}$ is the "average" operator for scalars,

$$\{\!\{\Pi\}\!\} = \frac{1}{2}\left(\Pi^+ + \Pi^-\right). \tag{2.106}$$

Note that there are no jumps in the horizontally aligned facets since $[\![\theta \boldsymbol{w}]\!] = 0$. When using the lowest-order Raviart-Thomas elements on quadrilateral-faced hexahedra (such as cubes), it can be shown that this discretisation reduces to the finite difference discretisation currently used in the Met Office Unified Model as described in Wood et al. (2014).

For equations (2.101) and (2.102), we can use the previously mentioned advection schemes for $\theta$ and $\rho$. For $\rho$, we can apply standard DG advection and we use the SUPG stabilisation scheme for $\theta$. This leads to the following compatible finite element formulation (semi-discrete): find $\boldsymbol{u} \in W_2$, $\rho, \Pi \in W_3$, and $\theta \in W_\theta$ such that

$$\begin{aligned}\int_\Omega \boldsymbol{w} \cdot \frac{\partial \boldsymbol{u}}{\partial t}\,\mathrm{d}\boldsymbol{x} + \int_\Omega \boldsymbol{w} \cdot ((\boldsymbol{u} \cdot \nabla)\boldsymbol{u})\,\mathrm{d}\boldsymbol{x} + \int_\Omega \boldsymbol{w} \cdot 2\boldsymbol{\Omega} \times \boldsymbol{u}\,\mathrm{d}\boldsymbol{x} \\ - \int_\Omega c_p \Pi \nabla_h \cdot (\theta \boldsymbol{w})\,\mathrm{d}\boldsymbol{x} + \int_{\mathcal{E}_z^\circ} c_p [\![\theta \boldsymbol{w}]\!] \{\!\{\Pi\}\!\}\,\mathrm{d}S \\ + \int_\Omega g \boldsymbol{w} \cdot \hat{\boldsymbol{z}}\,\mathrm{d}\boldsymbol{x} = 0,\end{aligned} \tag{2.107}$$

$$\begin{aligned}\int_\Omega (\gamma + \alpha \Delta z(\hat{\boldsymbol{w}} \cdot \nabla)\gamma) \frac{\partial \theta}{\partial t}\,\mathrm{d}\boldsymbol{x} + \int_\Omega (\gamma + \alpha \Delta z(\hat{\boldsymbol{w}} \cdot \nabla)\gamma)\, \boldsymbol{u} \cdot \nabla \theta\,\mathrm{d}\boldsymbol{x} \\ + \int_{\mathcal{E}_z^\circ} [\![(\gamma + \alpha \Delta z(\hat{\boldsymbol{w}} \cdot \nabla)\gamma)\boldsymbol{u}]\!]\widetilde{\theta} - [\![(\gamma + \alpha \Delta z(\hat{\boldsymbol{w}} \cdot \nabla)\gamma)\boldsymbol{u}\theta]\!]\,\mathrm{d}S = 0,\end{aligned} \tag{2.108}$$

$$\int_\Omega \phi \frac{\partial \rho}{\partial t}\,\mathrm{d}\boldsymbol{x} - \int_\Omega \phi (\boldsymbol{u} \cdot \nabla_h)\rho\,\mathrm{d}\boldsymbol{x} + \int_{\mathcal{E}^\circ} [\![\phi \boldsymbol{u}]\!]\widetilde{\rho}\,\mathrm{d}S = 0, \tag{2.109}$$

for all $\boldsymbol{w} \in W_2$, $\phi \in W_3$, and $\gamma \in W_\theta$. One additional equation is required to close the system, so we use the Galerkin projection equation for the Exner pressure:

$$\int_\Omega \phi \Pi\,\mathrm{d}\boldsymbol{x} - \int_\Omega \phi \left(\frac{R_d \rho \theta}{p_0}\right)^{\frac{\kappa}{1-\kappa}} \mathrm{d}\boldsymbol{x} = 0 \quad \text{for all } \phi \in W_3. \tag{2.110}$$

Equations (2.107)–(2.110) make up the compatible finite element formulation of the three-dimensional compressible Euler equations. The properties of the discretisation, as well as numerical studies, are still currently undergoing research.

# Chapter 3
# Firedrake

The Firedrake project is an automated system for solving variational problems using the finite element method (Rathgeber et al., 2016). "Automated" in this context means that the user specifies the finite element problem symbolically using high level code that reflects the mathematical description of the variational form. The high performance implementation of the assembly operations for the discrete linear and bilinear operators is then generated by a sequence of specialised compiler passes which apply symbolic mathematical transformations to the input equations to ultimately produce C (and C++) code. Firedrake compiles this code and executes it to create a linear system, which is solved by PETSc (Balay et al., 2018a, 2018b, 1997). Nonlinear systems can be automatically linearised and solved with a Newton-like method, or the user can specify a linearisation symbolically and code their own Picard iteration or similar.

In comparison with conventional finite element libraries, and even more so with hand written code, Firedrake provides a much higher productivity mechanism for solving finite element problems while simultaneously applying sophisticated performance optimisations that few users would have the resources to code by hand. Firedrake builds on the concepts and some of the code of the FEniCS project (Logg et al., 2012) and also shares similarities with FreeFEM++ (Hecht, 2012). In contrast with both of those projects, Firedrake has a number of features which are required in order to effectively employ compatible finite element formulations for geophysical fluid dynamics. Of particular importance in this regard are Firedrake's support for extruded meshes, a full range of tensor product elements, and composable operator-aware preconditioners.

## 3.1 A mixed Poisson Example

Consider the homogeneous Dirichlet problem for the two-dimensional Poisson equation,



T. H. Gibson et al., *Compatible Finite Element Methods for Geophysical Flows*, Mathematics of Planet Earth, https://doi.org/10.1007/978-3-030-23957-2_3

$$-\nabla \cdot \nabla u = f \quad \text{in } \Omega, \tag{3.1}$$
$$u = 0 \quad \text{on } \partial\Omega, \tag{3.2}$$

where $\Omega = [0,1]^2$ is the unit square and $f(x,y) = -2(x-1)x - 2(y-1)y$. This has an analytic solution $u(x,y) = x(1-x)y(1-y)$. By introducing the flux $\boldsymbol{\sigma} = -\nabla u$, we can reformulate the problem above as the first-order system of equations

$$\boldsymbol{\sigma} + \nabla u = 0 \quad \text{in } \Omega, \tag{3.3}$$
$$\nabla \cdot \boldsymbol{\sigma} = f \quad \text{in } \Omega, \tag{3.4}$$
$$u = 0 \quad \text{on } \partial\Omega. \tag{3.5}$$

The corresponding weak formulation of the mixed system above is posed as follows: find $(\boldsymbol{\sigma}, u) \in H(\text{div}) \times L^2$ such that

$$\int_\Omega \boldsymbol{\sigma} \cdot \boldsymbol{\tau} \, \mathrm{d}\boldsymbol{x} - \int_\Omega \nabla \cdot \boldsymbol{\tau} u \, \mathrm{d}\boldsymbol{x} = 0, \tag{3.6}$$
$$\int_\Omega v \nabla \cdot \boldsymbol{\sigma} \, \mathrm{d}\boldsymbol{x} = \int_\Omega f v \, \mathrm{d}\boldsymbol{x}, \tag{3.7}$$

for all test functions $(\boldsymbol{\tau}, v) \in H(\text{div}) \times L^2$. One option would be to seek $\boldsymbol{\sigma} \in [H^1]^2$ using the space of continuous piece-wise linear polynomials, and seek $u$ in the space of piece-wise constant functions. Mathematically, finding $\boldsymbol{\sigma} \in [H^1]^2$ is sound, as $H^1 \subset H(\text{div})$. However, this method is provably unstable. The simplest stable pair involves using the lowest-order Raviart-Thomas (RT) triangular elements (Raviart and Thomas, 1977) for $\boldsymbol{\sigma}$, and the space of piece-wise constants for $u$. At lowest-order, the RT space consists of vector-valued functions whose normal component on each edge of a triangular cell are constant, and continuous between adjacent cells. This space is a subspace of $H(\text{div})$, but not of $H^1$. The difference between the stable and unstable methods are illustrated in Figure 3.1.

Let's return to equations (3.6) and (3.7), and consider the computation behind Figure 3.1b. Let $\Omega$ be a triangulation of the unit square, let $U$ be the lowest order Raviart-Thomas element on $\Omega$, and let $V$ be the space of piecewise constant functions on $\Omega$. The discrete variational problem is then: find $(\boldsymbol{\sigma}, u) \in U \times V$ such that

$$\int_\Omega \boldsymbol{\sigma} \cdot \boldsymbol{\tau} \, \mathrm{d}\boldsymbol{x} - \int_\Omega u \nabla \cdot \boldsymbol{\tau} \, \mathrm{d}\boldsymbol{x} = 0, \tag{3.8}$$
$$\int_\Omega \nabla \cdot \boldsymbol{\sigma} v \, \mathrm{d}\boldsymbol{x} = \int_\Omega f v \, \mathrm{d}\boldsymbol{x}, \tag{3.9}$$

for all test functions $(\boldsymbol{\tau}, v) \in U \times V$. Firedrake user code is written in Python, so the first step is to import the Firedrake module:

```
from firedrake import *
```

Next we need a mesh. For simple domains such as the unit square, Firedrake provides built-in meshing functions:

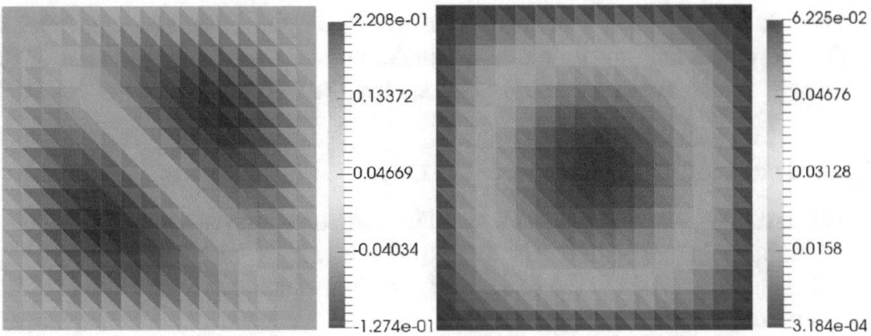


(a) The unstable method. (b) Lowest-order RT method (stable).

**Fig. 3.1** The results of using an unstable and stable discretisation of equations (3.6) and (3.7). In both cases, $u$ lies in the space of piece-wise constant functions. In the unstable case, $\sigma$ lies in the space of continuous piece-wise linear vector polynomials, leading to an oscillatory solution in $u$; in the stable method, $\sigma$ is in the lowest-order RT space, leading to an accurate approximation (up to discretisation error) of the true solution

```
mesh = UnitSquareMesh(16, 16)
```

We also need the function spaces $U$ and $V$. This is achieved by associating the mesh with the relevant finite element:

```
U = FunctionSpace(mesh, "RT", 1)
V = FunctionSpace(mesh, "DG", 0)
```

$U$, and $V$ are now symbolic variables representing function spaces. They also contain the computational implementation of the function space, recording the association of degrees of freedom with the mesh and pointing to the finite element basis, but the user does not usually need to pay any attention to this: the function space just behaves as a mathematical object. Since function spaces are mathematical symbols, we can combine them in the natural way to create the mixed function space on which the problem is posed:

```
W = U * V
```

The Firedrake user writes the variational problem symbolically in mathematical notation. We therefore need the symbols for the trial functions $(\sigma, u)$ and test functions $(\tau, v)$:

```
sigma, u = TrialFunctions(W)
tau, v = TestFunctions(W)
```

The other symbol in the variational problem is the forcing function $f$. There are various choices for the mathematical representation of $f$, but we will represent it as a function in the finite element space $V$:

```
f = Function(V)
```

The values that $f$ takes over the domain are given by an explicit formula in terms of the coordinate field of the mesh. Firedrake provides access to symbols for the coordinates which can then be used to interpolate the field values at the nodes of $f$:

```
x, y = SpatialCoordinate(mesh)
f.interpolate(-2*(x-1)*x - 2*(y-1)*y)
```

This last line is the first point at which Firedrake transforms a symbolic operation into a numerical computation. The `interpolate` method generates *C* code which evaluates this expression at the nodes of $f$, and immediately executes it to populate the values of $f$.

We are now in a position to define our variational problem:

```
a = (dot(tau, sigma) - u*div(tau) + div(sigma)*v)*dx
L = f*v*dx
```

observe that this is simply a plain text representation of (3.8) and (3.9) with the integral over the domain represented by multiplication by the measure $\mathrm{d}x$. This representation of variational problems is called the Unified Form Language (UFL) and was created for the FEniCS project (Alnæs et al., 2014). Firedrake adopts UFL both because it provides a powerful mathematical mechanism for specifying variational problems, and to provide users with a largely identical user interface between Firedrake and FEniCS.

We finally now come to solve the variational problem. We do this by defining a function in the solution space $W$ to hold the solution, and simply instructing Firedrake to solve the system:

```
w = Function(W, name="Solution")
solve(a == L, w)
```

The last line once again converts symbolic mathematics into a computation. Firedrake's compiler, TSFC (Homolya et al., 2018), generates the code which assembles the sparse matrix for the left hand side and the vector on the right. This code is then executed to create the matrix and vector, and the resulting system is passed to PETSc for solution. As we shall see later, the solution process is also fully programmable, enabling the creation of sophisticated solvers by composing together multiple layers of Krylov methods and preconditioners such as geometric or algebraic multigrid.

## 3.2 Extruded Meshes

The domains in which geophysical flows occur are typically a thousand times wider than they are deep. For instance the troposphere is approximately 10 km deep while the circumference of the Earth is around 40,000 km. Similarly, ocean basins are typically thousands of metres deep, but thousands of kilometres across. In addition to the domain anisotropy, the orientation of the gravitational acceleration, and hence of stratification, results in a large scale and process separation between the vertical and

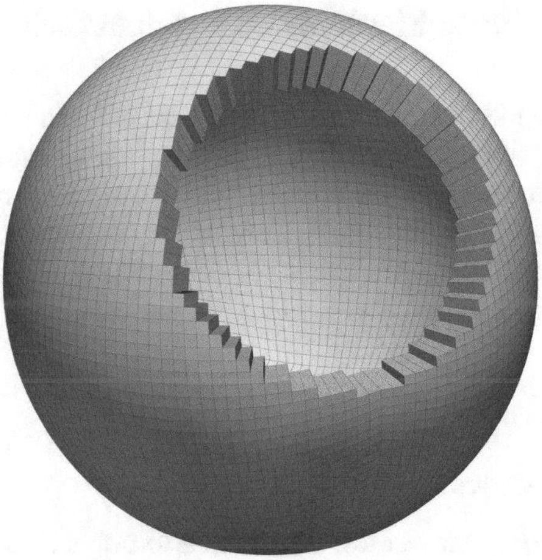

**Fig. 3.2** An extruded mesh of the cubed sphere with $32 \times 32$ quadrilaterals per cube face (refinement level 5) and 20 layers

horizontal directions. This separation motivates the near-universal choice of meshes which, while they may be structured or unstructured in the horizontal dimension, are composed of vertically aligned layers of cells. Firedrake supports this using the concept of an extruded mesh. An extruded mesh is constructed by taking a one- or two-dimensional base mesh and replicating it as a series of connected layers.

Firedrake attempts to reflect the mathematical structure of the problem studied, so an extruded mesh is formed by applying an extrusion operation to a base mesh. In the case of Figure 3.2, that is:

```
base_mesh = CubedSphereMesh(radius=6400.e6, refinement_level=5,
                            degree=2)
mesh = ExtrudedMesh(base_mesh, layers=20, layer_height=10,
                    extrusion_type="radial")
```

This creates an extruded cubed sphere mesh with $2^5$ quadrilateral cells along each cube edge and using biquadratic quadrilaterals to approximate the curved surface of the sphere. The construction of manifold meshes, such as the surface of a sphere, is facilitated by the technology developed in Rognes et al. (2013). There are 20 layers, each of which is 10 m high. Many other variations are possible, including variable layer height and different numbers of layers in different parts of the domain. The latter is particularly useful for ocean simulations.

## 3.3 Compatible Finite Elements on Product Cells

The cells of extruded meshes are a product of a base cell and a vertical interval cells. The construction of finite element approximations on extruded meshes therefore demands the ability to specify finite element spaces on these *product cells*. Firedrake supports the construction product finite element spaces, which may be manipulated to give the desired continuity properties for constructing $H^1$, $H(\mathrm{curl})$, $H(\mathrm{div})$, and $L^2$ spaces on extruded meshes. The implementation details are discussed in McRae et al. (2016).

As we have seen before, creating a function space on non-extruded meshes uses the following syntax:

```
V = FunctionSpace(mesh, "CG", 1)
```

While this sufficient for many cases, Firedrake also offers a more flexible means of instantiating more complex function spaces. Using the more general syntax:

```
V = FunctionSpace(mesh, element)
```

where `element` is a UFL `FiniteElement` Python object, we can construct more involved finite element spaces by generating and manipulating existing `FiniteElement` objects. The will enable us to generate finite elements on product cells by taking tensor products of individual finite elements with other element instances.

### 3.3.1 Product Spaces in Two Dimensions

In two dimensions, we consider the case of an extruded mesh constructed by taking the product of two intervals, yielding a structured quadrilateral mesh. Finite elements on intervals are scalar-valued and are either $H^1$ or $L^2$, depending on nodal continuity. A product mesh, say a unit square consisting $20 \times 20$ structured quadrilaterals, can be constructed in Firedrake with:

```
N = 20
base_mesh = UnitIntervalMesh(N)
mesh = ExtrudedMesh(base_mesh, layers=N, layer_height=1./N)
```

We first consider two scalar-valued elements in two dimensions, namely the $H^1$ and $L^2$ finite element spaces. For $H^1$, the elements must have nodes associated with the vertices, which enforces global continuity between neighbouring cells. In Firedrake, this is accomplished by performing algebra on UFL finite elements of the appropriate type. In this case, two $H^1$ elements $P_2$ and $P_3$ defined on the interval, of degrees 2 and 3 respectively, produce the desired $H^1$ space by simply taking their product and defining a new function space on the extruded interval mesh:

```
P2 = FiniteElement("CG", interval, 2)
P3 = FiniteElement("CG", interval, 3)
```

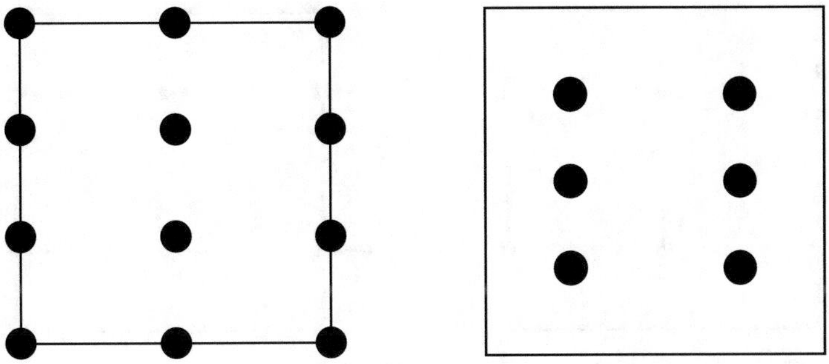

(a) The result of taking the product of $P_2$ with $P_3$.

(b) The result of taking the product of $dP_1$ with $dP_2$.

**Fig. 3.3** Finite element diagrams for the scalar elements in $H^1$ (Figure 3.3a) and $L^2$ (Figure 3.3b) on a product cell consisting of two intervals

```
H1_element = TensorProductElement(P2, P3)
H1 = FunctionSpace(mesh, H1_element)
```

Here, we use the `TensorProductElement` operator, which takes a two defined `FiniteElement` objects and produces a new element on the product cell. This operator is the basis for the construction of many finite elements on extruded meshes, including those defining div- or curl-conforming spaces.

The discontinuous $L^2$ product element must only have nodes associated with cell interiors. Therefore, the constituent interval finite elements must also have nodes only associated with their interiors. As before, this is done by taking two $L^2$ interval elements, say $dP_1$ and $dP_2$, and forming the new space on their product:

```
dP1 = FiniteElement("DG", interval, 1)
dP2 = FiniteElement("DG", interval, 2)
L2_element = TensorProductElement(dP1, dP2)
L2 = FunctionSpace(mesh, L2_element)
```

The resulting product elements for both scalar $H^1$ and $L^2$ are illustrated in Figure 3.3.

The vector-valued $H(\text{curl})$ and $H(\text{div})$ spaces require not just the correct association between nodes and topology, but also the correct pull-back to the reference element. For a more general discussion on the generation of $H(\text{curl})/H(\text{div})$ finite element spaces and their associated pull-backs, we refer the reader to (Rognes et al. 2009). Starting with $H(\text{curl})$, the element on the product cell must have nodes associated with the edges of the reference cell. In Firedrake, taking the product of an interval's vertex with an interval's interior nodes produces an element with nodes on the vertically aligned edges (the left and right rides of a quadrilateral). That is, we take the product of an $H^1$ element with an $L^2$ element:

```
S = TensorProductElement(P2, dP1)
```

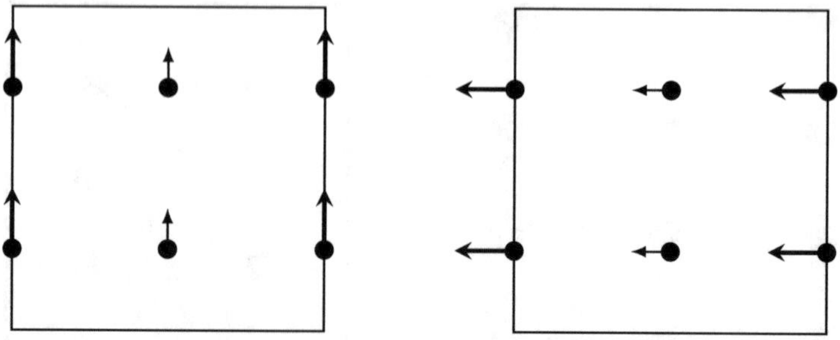

(a) The curl element with tangential nodes on the vertically-aligned edges of the quadrilateral.

(b) The div element with normal component pointing in the $-\boldsymbol{e}_0$ direction on the vertically-aligned edges.

**Fig. 3.4** Finite element diagrams for the vector elements in $H(\mathrm{curl})$ (Figure 3.4a) and $H(\mathrm{div})$ (Figure 3.4b) on a product cell consisting of two intervals

Now we need to transform this scalar-valued element into a vector $H(\mathrm{curl})$ element. To do this, we need to multiply by the vector $\boldsymbol{e}_1 = (0, 1)^T$, and set the pull-back to be covariant Piola rather than affine. This ensures that we preserve the tangential components when mapping from physical to reference element. The result in Firedrake is simply:

```
Hcurl = FunctionSpace(mesh, HCurl(S))
```

The operator `HCurl` forms the correct vector element from the product of the two interval elements. For $H(\mathrm{div})$, we construct it in the same manner as before, except we instead multiply by the vector $-\boldsymbol{e}_0 = (-1, 0)^T$, where the minus sign is useful for consistent orientation. The pull-back must now be contravariant Piola, which preserves the normal components when mapped from physical to reference element. We simply use the `HDiv` operator on the scalar product element:

```
Hdiv = FunctionSpace(mesh, HDiv(S))
```

See Figure 3.4 for an illustration of the resulting $H(\mathrm{curl})$ and $H(\mathrm{div})$ elements on the extruded interval mesh.

### 3.3.2 Product Spaces in Three Dimensions

For full three-dimensional formulations on extruded meshes, the construction of compatible finite element spaces is similar to that of the two-dimensional case. First, we need to define appropriate elements on the horizontal "base" mesh, followed by spaces on the vertical one-dimensional interval mesh. Consider the case where a horizontal triangular (structured) mesh (constructed by dividing each cell of a $10 \times 10$ quadrilateral mesh into two triangular cells of equal area) is extruded upwards by 10 levels:

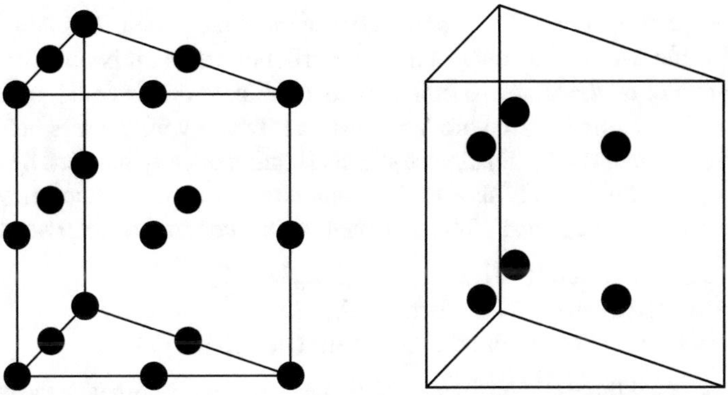

(a) The result of taking the product of continuous quadratic elements on triangles and intervals.

(b) The result of taking the product of discontinuous elements of linear degree on triangles and intervals.

**Fig. 3.5** Finite element diagrams for the scalar elements in $H^1$ (Figure 3.5a) and $L^2$ (Figure 3.5b) on triangular prisms

```
N = 10
base_mesh = UnitSquareMesh(N, N)
mesh = ExtrudedMesh(base_mesh, layers=N, layer_height=1./N)
```

This produces a mesh of the unit cube tessellated by triangular prism cells. As in the two-dimensional case, forming the $H^1$ and $L^2$ finite element spaces simply consists of taking products of $H^1$ and $L^2$ elements respectively (see Figure 3.5 for an illustration):

```
P2t = FiniteElement("CG", triangle, 2)
P2i = FiniteElement("CG", interval, 2)
H1_element = TensorProductElement(P2t, P2i)
H1 = FunctionSpace(mesh, H1_element)

dP1t = FiniteElement("DG", triangle, 1)
dP1i = FiniteElement("DG", interval, 1)
L2_element = TensorProductElement(dP1t, dP1i)
L2 = FunctionSpace(mesh, L2_element)
```

The curl-conforming element, as before, must have nodes topologically associated with cell edges. We shall construct a full $H(\texttt{curl})$ space by constructing the appropriate horizontally- and vertically-aligned vector-valued spaces. In both cases, an appropriate Piola transform is needed to map from physical to reference cell.

To construct the horizontal curl-conforming element, one may take the product of either an $H(\text{div})$ or $H(\text{curl})$ element with nodes on the edges of a triangle with an $H^1$ interval element. This produces a vector-valued element with nodes on the horizontally-aligned triangular bases of the product cell. This product naturally takes values in $\mathbb{R}^2$, however curl-conforming elements in three dimensions must take values in $\mathbb{R}^3$. If the two-dimensional element is curl-conforming, it is sufficient to interpret the product as the first two components of a three-dimensional vector. If the element is div-conforming, the product must be rotated by 90 degrees before transforming into a three-dimensional vector. This is automatically handled by the built-in `HCurl` operator previously shown. For simplicity, we create our element using the product of a two-dimensional $H(\text{curl})$ element with a continuous interval element:

```
N2_1 = FiniteElement("N2curl", triangle, 1)
P2i = FiniteElement("CG", interval, 2)
Hcurl_h = HCurl(TensorProductElement(N2_1, P2i))
```

For the vertical three-dimensional curl element, we can simply take the product of a continuous triangular element with a discontinuous interval element. This produces a scalar-valued element with nodes on the vertically-aligned quadrilateral faces of the triangular prism. Multiplying by the basis vector $\boldsymbol{e}_2 = (0,0,1)^T$ yields a vector-valued element whose tangential components are continuous across all edges and faces:

```
Hcurl_v = HCurl(TensorProductElement(P2t, dP1i))
```

With both the horizontal and vertical parts of the three-dimensional curl element, we can construct the full space by taking the direct sum of both parts. In Firedrake, this is known as an "enriched element" whose nodes are a concatenation of the nodes of the constituent sub-elements

```
Hcurl_element = Hcurl_h + Hcurl_v
Hcurl = FunctionSpace(mesh, Hcurl_element)
```

For a visualisation of the horizontal and vertical components of the resulting curl-conforming element in $\mathbb{R}^3$, see Figure 3.6

Finally, the construction of the $H(\text{div})$ element on triangular prisms follows almost identically from the curl-conforming example. To construct the full space, we need to define both the horizontal and vertical parts of the element and an appropriate Piola transform. The horizontal space, as with $H(\text{curl})$, the horizontal part can be constructed from a two-dimensional div-conforming element (i.e. Raviart-Thomas) and a discontinuous interval element. The vertical part can be constructed by transforming the scalar-valued product of an $H^1$ triangular element with an $L^2$ interval element into a vector-valued one via multiplying by $\boldsymbol{e}_2$. The full space (see Figure 3.7 for diagrams on the horizontal and vertical parts) is constructed as before:

```
RT2 = FiniteElement("RT", triangle, 2)
Hdiv_h = HDiv(TensorProductElement(RT2, dP1i))
Hdiv_v = HDiv(TensorProductElement(dP1t, P2i))
Hdiv_element = Hdiv_h + Hdiv_v
Hdiv = FunctionSpace(mesh, Hdiv_element)
```

Having constructed the required function spaces on extruded meshes, the formulation of finite element problems is nearly identical to the non-extruded case. Several examples will be shown in Chapter 5. Next, we discuss Firedrake's infrastructure for constructing sophisticated solvers.

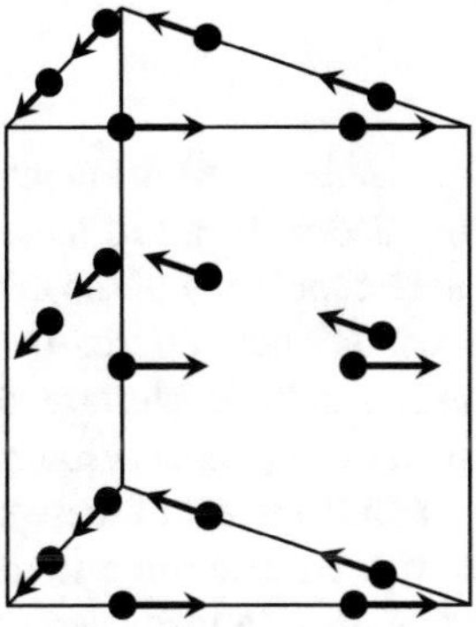

(a) The result of taking the product the Nédélec (2nd kind) element on triangles with an $H^1$ quadratic interval element. The nodes are transformed to support curl-conforming evaluation.

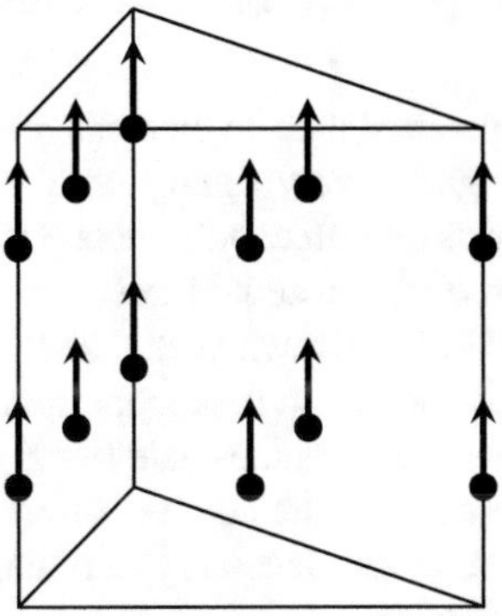

(b) The result of taking the product of an $H^1$ element of degree 1 on triangles with an $L^2$ quadratic element on an interval. The nodes are transformed to support curl-conforming evaluation.

**Fig. 3.6** Finite element diagrams for the vector-valued horizontal (Figure 3.6a) and and vertical (Figure 3.6b) components of the curl-conforming space on triangular prisms

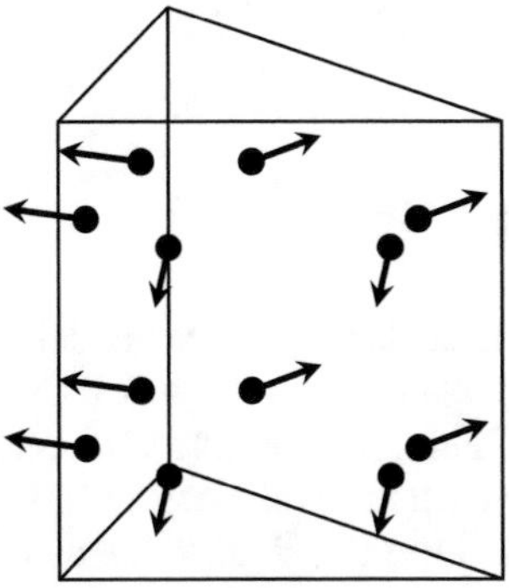

(a) The result of taking the product of a 2nd order Raviart-Thomas element on triangles with an $H^1$ quadratic interval element. The nodes are transformed to support div-conforming evaluation.

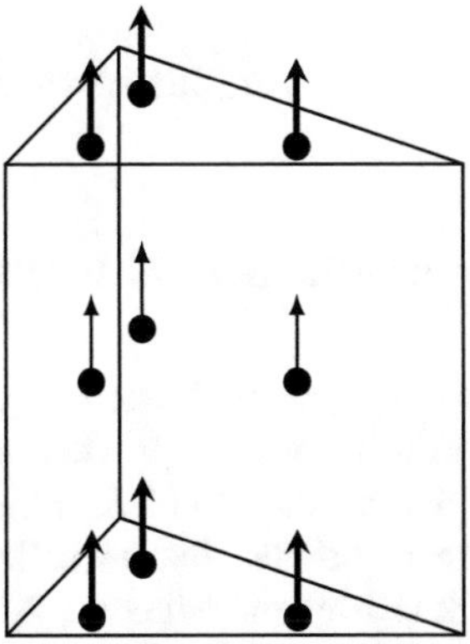

(b) The result of taking the product of an $H^1$ element of degree 1 on triangles with an $L^2$ quadratic element on an interval. The nodes are transformed to support div-conforming evaluation.

**Fig. 3.7** Finite element diagrams for the vector-valued horizontal (Figure 3.7a) and and vertical (Figure 3.7b) components of the div-conforming space on triangular prisms

## 3.4 Composable Preconditioners

Having specified a finite element problem, we then need to solve it somehow. Firedrake integrates very tightly with PETSc, offering access to a full suite of algebraic solvers and preconditioners. Of particular interest for many of the mixed finite element models presented later are the block preconditioners offered by PETSc's PCFIELDSPLIT (Brown et al., 2012). These present an algebraic interface to building many well-known block preconditioners, such as Schur complement eliminations, or block relaxation schemes (see Benzi et al. 2005) for an overview of these schemes). Firedrake extends the support offered by PETSc with an extensible framework to provide the (often necessary) auxiliary operators that arise in many such preconditioning schemes.

Firedrake's philosophy when specifying solvers, follows that of PETSc: it should be possible to configure the solver independently of the discretisation and problem setup. This enables using robust direct solvers for small problems when developing the model discretisation, followed by selecting scalable solvers when the discretisation is set up. We therefore always configure solvers using PETSc options: all solvers accept a dictionary of `solver_parameters` which are passed directly to PETSc to setup the solver. For example, to configure a solver to use the method of conjugate gradients, we write, in addition to the other arguments defining the problem to be solved:

```
solve(..., solver_parameters={"ksp_type": "cg"})
```

### 3.4.1 Preconditioners with Auxiliary Operators

For many block preconditioners, in particular those that involve Schur complements, it is useful to be able to provide an *auxiliary operator* to the solver that is not a block in the original problem. Firedrake facilitates this by allowing one to write Python preconditioners that are called by PETSc but have full access to the Firedrake framework (Kirby and Mitchell, 2018). We will use this feature extensively in the example hybridised solver of Sections 5.3 and 4.5.2, and introduce it here with a simpler example: solving Stokes' equations.

Our aim is to find $(\boldsymbol{u}, p) \in U \times V$ such that

$$\nu \int_\Omega \nabla \boldsymbol{u} : \nabla \boldsymbol{v} \, \mathrm{d}\boldsymbol{x} - \int_\Omega p \nabla \cdot \boldsymbol{v} \, \mathrm{d}\boldsymbol{x} = \int_\Omega \boldsymbol{f} \cdot \boldsymbol{v} \, \mathrm{d}\boldsymbol{x}, \tag{3.10}$$

$$- \int_\Omega \nabla \cdot \boldsymbol{u} q \, \mathrm{d}\boldsymbol{x} = 0 \tag{3.11}$$

$$\tag{3.12}$$

for all test functions $(\boldsymbol{v}, q) \in U \times V$, where $\nu$ is the viscosity. We will solve a lid-driven cavity problem in a unit square domain, so we have $\boldsymbol{f} = 0$, and boundary conditions

$$\boldsymbol{u} = \begin{pmatrix} \frac{x^2(2-x)^2y^2}{4} \\ 0 \end{pmatrix} \qquad \text{on } \Gamma_1 = \{y = 1\} \tag{3.13}$$

$$\boldsymbol{u} = 0 \qquad \text{otherwise} \tag{3.14}$$

along with homogeneous Neumann conditions for the pressure variable.

As for the simple example of Section 3.1, we start by importing Firedrake and defining a mesh:

```
from firedrake import *
mesh = UnitSquareMesh(64, 64)
```

We will also need the function spaces $U$ and $V$ for which we use the Taylor-Hood pair $P_2 - P_1$.

```
V = VectorFunctionSpace(mesh, "CG", 2)
Q = FunctionSpace(mesh, "CG", 1)
W = V*Q
```

Although this problem is linear, we will set it up as a nonlinear problem, to show the difference. We therefore do not need any trial functions, but only test functions:

```
v, q = TestFunctions(W)
```

We will also need a `Function` to store the solution, and a symbolic representation of the two parts (the velocity and pressure), obtained with `split`:

```
w = Function(W)
u, p = split(w)
```

The viscosity is represented as a `Constant`

```
nu = Constant(0.0001)
```

although it could also be a `Function` if we wanted spatial variation or values from an external source. We are now ready to define the variational problem. All nonlinear problems in Firedrake are specified in residual form $F(u; v) = 0$. We therefore collect all terms in the residual together

```
F = nu*inner(grad(u), grad(v))*dx - p*div(v)*dx - div(u)*q*dx
```

Now we move on to the boundary conditions, the forcing vector can be easily written as a UFL expression. First we obtain symbolic expressions for the $x$ and $y$ coordinates in the physical mesh:

```
x, y = SpatialCoordinate(mesh)
```

Then we create a vector holding representing the forcing of eq. (3.13)

```
forcing = as_vector([0.25 * x**2 * (2-x)**2 *y**2, 0])
```

A strong, or Dirichlet, boundary condition is created by constructing a Python `DirichletBC` object, where we must provide the function space the condition applies to, the value, and the part of the mesh at which it applies. The latter uses integer *mesh markers* which are used by most mesh generation software to tag entities of meshes. The builtin meshes (such as the `UnitSquareMesh` we are using here) automatically tag the boundary. For this mesh the tag 4 corresponds to the plane $y = 1$, the other three sides are labelled 1, 2, and 3. The solve call can handle both a single boundary condition, or, as is this case here, a list thereof:

```
bcs = [DirichletBC(W.sub(0), forcing, 4),
       DirichletBC(W.sub(0), zero(mesh.geometric_dimension()),
                   (1, 2, 3))]
```

Note how we are applying boundary conditions to the velocity part of the mixed finite element space $W$, and so indicate that with `W.sub(0)`. To apply conditions to the pressure space, we would use `W.sub(1)`.

Now we come to configure the solver. This problem has a null space of all functions of constant pressure, so we need to ensure that our solver removes this space. To do so, we build a null space object, which will subsequently be passed to the solver. PETSc will then take care of seeking a solution in the space orthogonal to the provided null space.

```
const_basis = VectorSpaceBasis(constant=True)
nullspace = MixedVectorSpaceBasis(W, [W.sub(0), const_basis])
```

Our solution strategy uses the block preconditioning scheme of (Silvester and Wathen 1994) in which the Schur complement is approximated by the inverse of a viscosity-weighted pressure mass matrix. We configure this solver by creating an appropriate dictionary of solver parameters. Since the problem is actually linear, although we have posed it as a nonlinear problem, we select PETSc's `ksponly` nonlinear solver (which just runs a linear solve) and configure appropriate tolerances for that linear solver, as well as asking PETSc to monitor the progress of the solve:

```
solver_parameters = {
    "snes_type": "ksponly",
    "ksp_rtol": 1e-8,
    "ksp_type": "gmres",
    "ksp_monitor": None,
```

The Schur complement block preconditioner is selected with:

```
    "pc_type": "fieldsplit",
    "pc_fieldsplit_type": "schur",
```

For this example we will use LU factorisation to invert the velocity block:

```
    "fieldsplit_0": {
        "ksp_type": "preonly",
        "pc_type": "lu",
    },
```

Notice how we can provide nested dictionaries to configure the subsolver. The inverse of the Schur complement is configured with the prefix `fieldsplit_1`. Unlike the velocity block, it is not explicitly formed, and therefore we need to provide an appropriate auxiliary operator to use in the preconditioner. Silvester and Wathen (1994) show that for Stokes' equations, the pressure Schur complement is spectrally equivalent to a pressure mass matrix, scaled by $-\nu^{-1}$, which can be assembled from the bilinear form:

$$-\int_{\Omega} \nu^{-1} p q \, \mathrm{d}\boldsymbol{x}. \tag{3.15}$$

This term does not appear in eq. (3.10), and hence we cannot construct it algebraically from the matrix we are trying to invert. Firedrake's solution to this is to support, with the help of PETSc, user-defined preconditioners that can provide the necessary auxiliary operators and inverse actions. To do so, we select a `python` preconditioner

```
    "fieldsplit_1": {
        "ksp_type": "preonly",
        "pc_type": "python",
```

and provide our own class as a constructor:

```
        "pc_python_type": "__main__.MassMatrix",
```

Finally, we must configure the mechanism for inverting the operator our class provides. Here, since the mass matrix is easy to invert, we use one application of incomplete LU factorisation.

```
        "mass_pc_type": "bjacobi",
        "mass_sub_pc_type": "ilu",
    }
}
```

It remains to define the `MassMatrix` class. Since preconditioning with an assembled, user-specified, operator is such a common requirement for block solvers, Firedrake provides a base class in which we only need to override the function returning the variational form that defines the operator.

```
class MassMatrix(AuxiliaryOperatorPC):
    _prefix = "mass_"
    def form(self, pc, test, trial):
        nu = self.get_appctx(pc)["nu"]
        a = -1/nu*inner(test, trial)*dx
        return (a, None)
```

The viscosity is extracted from a user-provided dictionary of state that is provided to the `solve` command. Finally, we are ready to solve the problem, we build the application context that contains the viscosity

```
appctx = {"nu": nu}
```

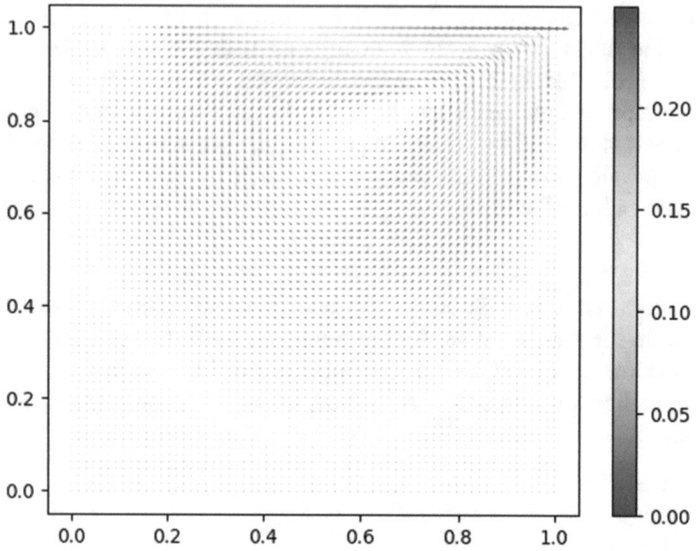


**Fig. 3.8** The resulting velocity field for the lid-driven cavity problem

and call solve passing the residual, and function to minimise the residual with respect to, boundary conditions, null space, the application context (containing $\nu$), and our dictionary of solver parameters.

```
solve(F == 0, w,
      bcs=bcs,
      nullspace=nullspace,
      appctx=appctx,
      solver_parameters=solver_parameters)
```

The resulting velocity field is visualised in Fig. 3.8. When the solve runs, the symbolic mathematics is converted into a computation, and PETSc is called to solve the resulting sparse linear systems. The difference from the example in Section 3.1 is that when building the preconditioner for the Schur complement, PETSc calls back into Firedrake to assemble to appropriate operator. This mutual callback interface significantly simplifies to construction of complex preconditioners, since we can separate the solver from the problem specification. For more details and examples, we refer to Kirby and Mitchell (2018).

# Chapter 4
# Models in Two-Dimensions

In this chapter we describe how to implement geophysical fluid dynamics models in Firedrake, focusing on two-dimensional models. The chapter is formed from code examples which highlight relevant aspects of the Firedrake modelling system. In particular, the definition and construction of compatible finite element spaces, the associated weak formulations and resulting discrete systems of PDEs.

In Section 4.1, we simulate the linear shallow water equations. In Section 4.2, we implement a discontinuous Galerkin method for tracer advection. Finally, in Section 4.3, we combine these aspects into a full nonlinear shallow water model.

## 4.1 Linear Shallow Water Equations

In Section 2.3.2, we derived a compatible finite element scheme for the linear shallow water equations, with an implicit midpoint timestepping scheme. Given a velocity field $u^n \in V_1$ and depth perturbation field $h^n \in V_2$ at time $t_n$, the fields at time $t_{n+1}$ satisfy

$$\int_\Omega w \cdot u^{n+1}\,\mathrm{d}x + \frac{\Delta t}{2}\int_\Omega w \cdot f\left(u^{n+1}\right)^{\perp}\,\mathrm{d}x \\ -\frac{g\Delta t}{2}\int_\Omega h^{n+1}\nabla\cdot w\,\mathrm{d}x = -R_u^n[w], \tag{4.1}$$

$$\frac{H\Delta t}{2}\int_\Omega \phi\nabla\cdot u^{n+1}\,\mathrm{d}x + \int_\Omega \phi h^{n+1}\,\mathrm{d}x = -R_h^n[\phi], \tag{4.2}$$

for all $w \in V_1$ and $\phi \in V_2$, where the right-hand side functionals are



T. H. Gibson et al., *Compatible Finite Element Methods for Geophysical Flows*, Mathematics of Planet Earth, https://doi.org/10.1007/978-3-030-23957-2_4

$$R_u^n[w] = \int_\Omega w \cdot u^n \, \mathrm{d}x - \frac{\Delta t}{2} \int_\Omega w \cdot f (u^n)^\perp \, \mathrm{d}x + \frac{g\Delta t}{2} \int_\Omega h^n \nabla \cdot w \, \mathrm{d}x, \tag{4.3}$$

$$R_h^n[\phi] = -\frac{H\Delta t}{2} \int_\Omega \phi \nabla \cdot u^n \, \mathrm{d}x + \int_\Omega \phi h^n \, \mathrm{d}x. \tag{4.4}$$

The physical parameters are $H$, the depth, $g$, the acceleration due to gravity, and $f$, the Coriolis parameter.

The fully discrete finite element problem produces a system of linear equations of the form

$$\begin{bmatrix} M_{1,f} & -\frac{g\Delta t}{2} D^T \\ \frac{H\Delta t}{2} D & M_2 \end{bmatrix} \begin{Bmatrix} U^{n+1} \\ H^{n+1} \end{Bmatrix} = - \begin{Bmatrix} R_{U^n} \\ R_{H^n} \end{Bmatrix}, \tag{4.5}$$

where $M_{1,f} = M_1 + \frac{\Delta t}{2} M_f$, $M_1$ and $M_2$ are mass matrices for $V_1$ and $V_2$, respectively, $M_f$ represents the contribution of the Coriolis force, $D$ represents the weak divergence, and $D^T$ the weak gradient. The linear system (4.5) defines a saddle-point system of equations, which is generally indefinite. This requires careful consideration when using iterative methods to invert the mixed system, many of which revolve around sophisticated preconditioners using Schur complement factorisations (Benzi et al., 2005). We shall return to this point in the next chapter, since iterative methods become essential in 3D problems. In this chapter, we will use direct methods like LU factorisations to solve the linear systems.

### 4.1.1 Example: Linear Solid Body Rotation

Using Firedrake to construct a finite element model requires a high-level specification of the finite element problem. Given a discretisation of the equations in weak form, i.e., (4.1)–(4.4), we assemble and solve the linear system described by (4.5) for the fields in the next time step. We repeat this until a specified final time. As a first example, we implement a simple solid body rotation test on the sphere, as described by Williamson et al. (1992).

First, we import Firedrake and create some relevant physical constants.

```
from firedrake import *

R0 = 6371220.0
R = Constant(R0)                # Radius of the earth [m]
H = Constant(5960.0)            # Mean depth [m]
day = 24. * 60. * 60.           # Seconds in a day [s]
Omega = Constant(7.292E-5)      # Angular rotation rate [rads]
g = Constant(9.80616)           # Accel. due to gravity [m/s^2]
```

With Firedrake imported, we now have access to all of its utility functions for setting up meshes, function spaces, and finite element weak forms. We now construct our domain, a triangulation of a sphere with radius $R_\odot = 6371220.0$ m.

```
9  mesh_degree = 3    # Cubic coordinate field
10 refinements = 5    # Number of refinements
11 mesh = IcosahedralSphereMesh(R0, refinements,
12                              degree=mesh_degree)
13 x = SpatialCoordinate(mesh)
14 global_normal = as_vector([x[0], x[1], x[2]])
15 mesh.init_cell_orientations(global_normal)
```

Here, we are using a cubic coordinate field to better resolve the curvature of the Earth. Our mesh is generated using five regular refinements of an icosahedron, resulting in a mesh consisting of 20,480 cells. In line 13, we extract the Cartesian coordinates and set up the orientation of the global normals on the sphere in lines 14 and 15. This is a necessary step in Firedrake when using $H(\text{div})$ vector fields on an immersed manifold such as the sphere or torus. We shall also use the coordinate field to construct explicit expressions for our Coriolis term and initial conditions.

Recall the two-dimensional finite element complexes outlined in Section 2.1. We now define appropriate finite element spaces for the velocity and depth fields. We choose $BDM_2$ for the velocity space $V_1$, and $P_1^{(DG)}$ for the depth space $V_2$.

```
16 model_degree = 2    # Degree of the finite element complex
17 V1 = FunctionSpace(mesh, "BDM", model_degree)
18 V2 = FunctionSpace(mesh, "DG", model_degree - 1)
19 V = V1 * V2    # Mixed space for velocity and depth
```

We declare our Coriolis term as a symbolic expression, then use this to construct a finite element function that lies in a continuous finite element space of the same degree as the coordinate field.

```
20 f_expr = 2 * Omega * x[2] / R
21 Vf = FunctionSpace(mesh, "CG", mesh_degree)
22 f = Function(Vf).interpolate(f_expr)    # Coriolis frequency
```

With our function spaces defined, we can now set up our initial conditions for the simulation. The solid body example starts with velocity and depth fields in geostrophic balance, given by the expressions

$$u^0 = \frac{u_{\max}}{R_\odot}(-y, x, 0), \quad h^0 = H - \left(R_\odot \Omega u_{\max} + \frac{u_{\max}^2}{2}\right)\frac{z^2}{gR_\odot^2}, \tag{4.6}$$

where $u_{\max} = 2\pi R_\odot/(12\text{days}) \approx 38.6\,\text{ms}^{-1}$ and $H = 5960.0\,\text{m}$. Since there is no forcing or topography, the solution should remain steady throughout the entire simulation. Using our previously defined fields and constants, we can concisely write out the initial conditions and create corresponding Firedrake `Function` objects.

```
23 u_0 = 2*pi*R0/(12*day)
24 u_max = Constant(u_0)  # Maximum amplitude of the zonal wind
25 u_exp = as_vector([-u_max*x[1]/R, u_max*x[0]/R, 0.0])
26 h_exp = H-((R*Omega*u_max+u_max*u_max/2.0)*(x[2]*x[2]/(R*R)))/g
```

```
27 u0 = Function(V1).project(u_exp)
28 h0 = Function(V2).interpolate(h_exp)
29 un = Function(V1).assign(u0)  # Fields at time-step n
30 hn = Function(V2).assign(h0)
```

Here, we store the initial conditions in functions to be used later on, as well as initialise the fields that we will update during the time integration. We now initialise the test and trial functions, and prepare time-stepping parameters.

```
31 tmax = 5 * day
32 Dt = 1000.0
33 dt = Constant(Dt)                          # Timestep size [s]
34 alpha = Constant(0.5)                      # Midpoint method
35 u, h = TrialFunctions(V)
36 w, phi = TestFunctions(V)
37 outward_normal = CellNormal(mesh)
38 perp = lambda u: cross(outward_normal, u)  # Perp operator
```

We run the simulation for a maximum of five days, using a timestep of $\Delta t =$ 1000s. To handle the Coriolis term, we also define the perp operator by taking cross products of velocities with cell normals. All of these are constructed using symbolic UFL operators. We are now ready to define the finite element forms for equations (4.1)–(4.4) in UFL:

```
39 uh_eqn = (inner(w, u) + alpha*dt*inner(w, f*perp(u))
40           - alpha*dt*g*div(w)*h
41           - inner(w, un)
42           + alpha*dt*inner(w, f*perp(un))
43           - alpha*dt*g*div(w)*hn
44           + phi*h + alpha*dt*H*phi*div(u)
45           - phi*hn + alpha*dt*H*phi*div(un))*dx
46 wn1 = Function(V)          # mixed func. for both fields (n+1)
47 un1, hn1 = wn1.split()     # Split func. for individual fields
48 uh_problem = LinearVariationalProblem(lhs(uh_eqn),
49                                       rhs(uh_eqn), wn1)
50 params = {'mat_type': 'aij',
51           'ksp_type': 'preonly',
52           'pc_type': 'lu',
53           'pc_factor_mat_solver_type': 'mumps'}
54 uh_solver = LinearVariationalSolver(uh_problem,
55                                     solver_parameters=params)
```

Lines 46–55 define the function for the fields at time step $n+1$, as well as the solver object for the linear system (4.5). The parameters in lines 50–53 communicate with PETSc, Firedrake's linear solver backend, to solve the linear system using the (sparse) LU algorithm from MUMPS (Amestoy et al., 2000). We specify `mat_type` to be `aij`, since PETSc defaults to a nested, or block, matrix type for assembled operators from mixed systems. The LU algorithm requires accessing all matrix entries via

$A = (a_{ij})_{i,j}$ rather than individual blocks. For block-preconditioning, a topic we will cover in a subsequent section, we will *not* use a monolithic `aij` matrix, which allows Firedrake and PETSc to efficiently extract individual operators that define the mixed matrix.

Now that the linear solver and relevant fields are defined, we are nearly ready to perform the time integration. We first write out the initial fields into a file which can be interpreted by software such as ParaView. We also define standard Python lists to keep track of relative errors in the prognostic fields, energy of the system, and write output.

```
# Write out initial fields
u_out = Function(V1, name="Velocity").assign(un)
h_out = Function(V2, name="Depth").assign(hn)
outfile = File("results/W2/linear_w2.pvd")
outfile.write(u_out, h_out)
t = 0.0
dumpfreq = 10    # Dump output every 10 steps
counter = 1
Uerrors = []
Herrors = []
t_array = []
# Energy functional
energy = 0.5*inner(un, un)*H*dx + 0.5*g*hn*hn*dx
energy_0 = assemble(energy)
energy_t = []
```

We now integrate forward in time, calculating diagnostics each timestep.

```
while t < tmax:  # Start time loop
    t += Dt
    t_array.append(t)
    uh_solver.solve()  # Solve for updated fields

    # Fields are solved in physical coordinates,
    # so we normalise errors by norm of the initial fields
    u0norm = norm(u0, norm_type="L2")
    h0norm = norm(h0, norm_type="L2")
    Uerr = errornorm(un1, un, norm_type="L2")/u0norm
    Herr = errornorm(hn1, hn, norm_type="L2")/h0norm
    Uerrors.append(Uerr)
    Herrors.append(Herr)

    # Use updated fields in the RHS for the next timestep
    un.assign(un1)
    hn.assign(hn1)

    # If energy is conserved, this should be close to 0
```

```
    energy_t.append((assemble(energy) - energy_0)/energy_0)

    counter += 1
    if counter > dumpfreq:
        u_out.assign(un)
        h_out.assign(hn)
        outfile.write(u_out, h_out)
        counter -= dumpfreq
```

There are a number of ways to interpret the results of the simulation. Using ParaView, we can directly read the pvd files and view the computed fields. Given that the exact solution is steady, we do not expect the numerical solution to deviate much from the initial conditions; this test really just serves as a form of verification that the code is a correct implementation. Figure 4.1 displays both the initial and final (day 5) fields computed during the simulation.

Using relative errors we computed in the example, we can get more concrete evidence that the flow remains steady. Figure 4.2 display the velocity and depth errors respectively. We can see that the errors in both fields oscillate, but remain under control. Factoring in the mesh resolution and the choice of discretisation, the errors

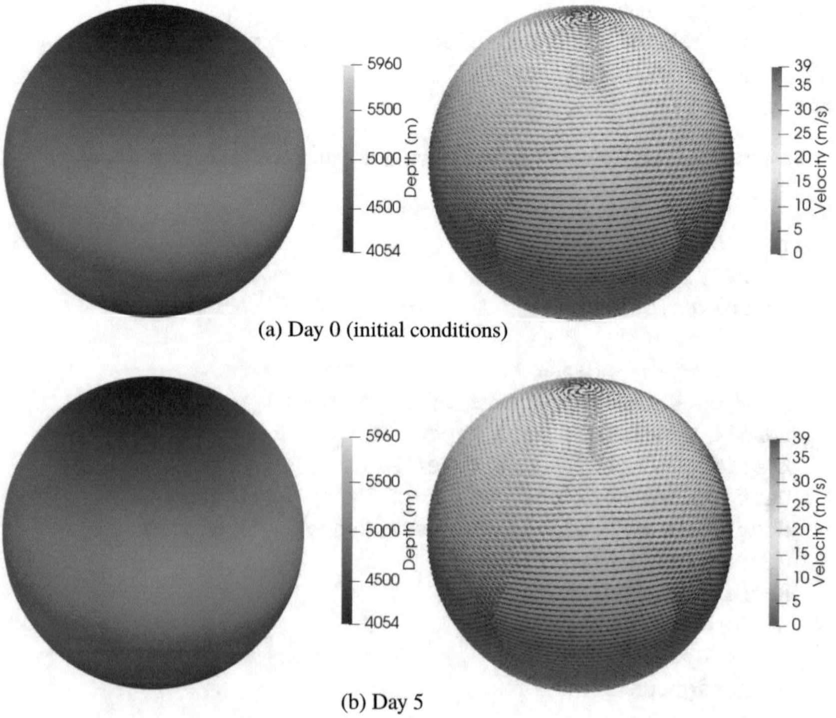


(a) Day 0 (initial conditions)

(b) Day 5

**Fig. 4.1** Solid body rotation example displaying the depth and velocity fields at day 0 (a) and day 5 (b)

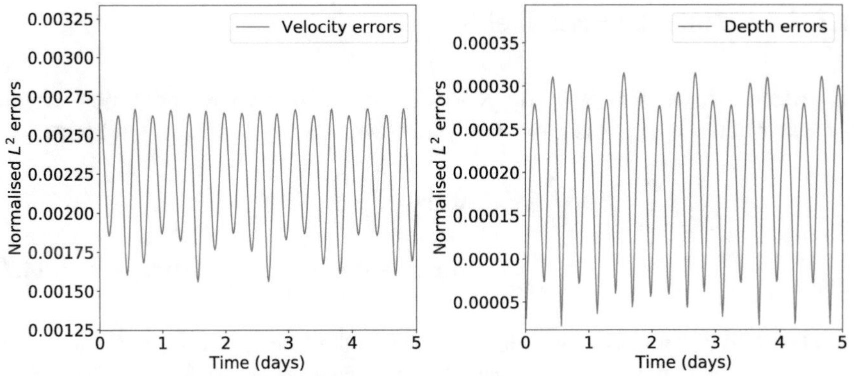


**Fig. 4.2** Velocity (left) and depth (right) errors over time

are relatively small considering the magnitude of the fields. Compatible finite element discretisations for shallow water schemes are discussed in Shipton et al. (2018), where a more detailed study is performed.

Recall in Section 2.3.1 the result on energy conservation when using a compatible finite element discretisation together with an energy conserving time-integrator (trapezoidal or implicit-midpoint rule). In our example, we compute the energy of the system (see line 68 and 69) at the start of the simulation, $E_0$. After updating the fields at time step $n$, we again evaluate the energy of the system $E_n$. If energy is conserved, we expect the relative difference $(E_n - E_0)/E_0$ to remain close to zero for all $n$. From the relative energy differences computed during the simulation, we can verify that our numerical scheme is conserving energy up to machine precision, shown in fig. 4.3.

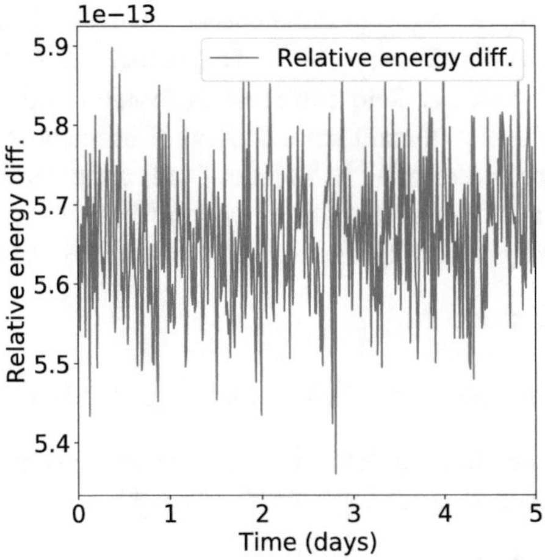


**Fig. 4.3** $(E_n - E_0)/E_0$ versus time (days) for the solid body rotation example

## 4.2 Advection of Scalar Fields

In Section 2.2, we presented an upwind discontinuous Galerkin advection scheme for scalar fields. If $q(x,t)$ is advected by a known velocity field $u(x,t)$, we have

$$\begin{aligned}\int_\Omega \phi \frac{\partial q}{\partial t}\,\mathrm{d}x = \int_\Omega q\nabla\cdot(\phi u)\,\mathrm{d}x - \int_\Gamma [\![\phi u]\!]\widetilde{q}\,\mathrm{d}S \\ - \int_{\partial\Omega_{\text{inflow}}} \phi q_{\text{in}} u\cdot n\,\mathrm{d}s - \int_{\partial\Omega_{\text{outflow}}} \phi q u\cdot n\,\mathrm{d}s, \end{aligned} \tag{4.7}$$

for all $\phi \in V$. $\widetilde{q}$ denotes the upwind value of $q$ on the facet, and $[\![v]\!]$ denotes $v^+\cdot n^+ + v^-\cdot n^-$, where $^+$ and $^-$ are arbitrary labels for the cells either side of a facet. As a timestepping scheme, we proposed a three-stage strong-stability-preserving Runge-Kutta (SSPRK) scheme (Shu and Osher, 1988). For an equation in the abstract form

$$\frac{\partial q}{\partial t} = \mathscr{L}(q), \tag{4.8}$$

the scheme is

$$q^{(1)} = q^n + \Delta t\mathscr{L}(q^n) \tag{4.9}$$

$$q^{(2)} = \frac{3}{4}q^n + \frac{1}{4}(q^{(1)} + \Delta t\mathscr{L}(q^{(1)})) \tag{4.10}$$

$$q^{n+1} = \frac{1}{3}q^n + \frac{2}{3}(q^{(2)} + \Delta t\mathscr{L}(q^{(2)})). \tag{4.11}$$

### 4.2.1 Example: Cosine-Bell–Cone–Slotted-Cylinder Test

In this worked example, we reproduce the classic cosine-bell–cone–slotted-cylinder advection test case from LeVeque (1996). The domain $\Omega$ is the unit square $\Omega = [0,1]\times[0,1]$, and the velocity field corresponds to solid body rotation $u = (0.5 - y, x - 0.5)$. Each side of the domain has a section of inflow and a section of outflow boundary. We therefore perform both the inflow and outflow integrals over the entire boundary, but construct them so that they only contribute in the correct places.

As usual, we start by importing Firedrake. We use a 128-by-128 mesh of squares.

```
from firedrake import *

mesh = UnitSquareMesh(128, 128, quadrilateral=True)
```

We set up a function space of discontinous bilinear elements for $q$, and a vector-valued continuous function space for our velocity field.

```
V = FunctionSpace(mesh, "DQ", 1)
W = VectorFunctionSpace(mesh, "CG", 1)
```

We set up the initial velocity field using a simple analytic expression.

```
x, y = SpatialCoordinate(mesh)

velocity = as_vector((0.5 - y, x - 0.5))
u = Function(W).interpolate(velocity)
```

Now, we set up the cosine-bell–cone–slotted-cylinder initial condition. The first four lines declare various parameters relating to the positions of these objects, while the analytic expressions appear in the last three lines.

```
bell_r0 = 0.15
bell_x0 = 0.25
bell_y0 = 0.5
cone_r0 = 0.15
cone_x0 = 0.5
cone_y0 = 0.25
cyl_r0 = 0.15
cyl_x0 = 0.5
cyl_y0 = 0.75
slot_left = 0.475
slot_right = 0.525
slot_top = 0.85

bell = 0.25*(1+cos(
    pi*min_value(sqrt(pow(x-bell_x0, 2) +
                      pow(y-bell_y0, 2))/bell_r0, 1.0))
)
cone = 1.0 - min_value(
    sqrt(pow(x-cone_x0, 2) + pow(y-cone_y0, 2))/cyl_r0, 1.0
)
slot_cyl = conditional(
    sqrt(pow(x-cyl_x0, 2) + pow(y-cyl_y0, 2)) < cyl_r0,
    conditional(And(And(x > slot_left,
                        x < slot_right), y < slot_top),
            0.0, 1.0), 0.0
)
```

We then declare the inital condition of $q$ to be the sum of these fields. Furthermore, we add 1 to this, so that the initial field lies between 1 and 2, rather than between 0 and 1. This ensures that we cannot get away with neglecting the inflow boundary condition. We also save the initial state so that we can check the $L^2$-norm error at the end.

```
q = Function(V).interpolate(1.0 + bell + cone + slot_cyl)
q_init = Function(V).assign(q)
```

We declare the output filename, and write out the initial condition.

```
outfile = File("results/DGadv.pvd")
outfile.write(q)
```

We will run for time $2\pi$, a full rotation. We take 1800 steps, giving a time step close to the CFL limit. We declare an extra variable `dtc`; for technical reasons, this means that Firedrake does not have to compile new C code if the user tries different time steps. Finally, we define the inflow boundary condition, $q_{\mathrm{in}}$. In general, this would be a `Function`, but here we just use a `Constant` value.

```
T = 2*pi
dt = T/1800.0
dtc = Constant(dt)
q_in = Constant(1.0)
```

Now we declare our variational forms. Solving for $\Delta q$ at each stage, the explicit time-stepping scheme means that the left hand side is just a mass matrix.

```
dq_trial = TrialFunction(V)
phi = TestFunction(V)
a = phi*dq_trial*dx
```

The right-hand-side is more interesting. We define `n` to be the built-in `FacetNormal` object; a unit normal vector that can be used in integrals over exterior and interior facets. We next define un to be an object which is equal to $u \cdot n$ if this is positive, and zero if this is negative. This will be useful in the upwind terms.

```
n = FacetNormal(mesh)
un = 0.5*(dot(u, n) + abs(dot(u, n)))
```

We now define our right-hand-side form `L1` as $\Delta t$ times the sum of four integrals.

The first integral is a straightforward cell integral of $q\nabla \cdot (\phi u)$. The second integral represents the inflow boundary condition. We only want this to contribute on the inflow part of the boundary, where $u \cdot n < 0$ (recall that $n$ is an outward-pointing normal). Where this is true, the condition gives the desired expression $\phi q_{\mathrm{in}} u \cdot n$, otherwise the condition gives zero. The third integral operates in a similar way to give the outflow boundary condition. The last integral represents the integral $\widetilde{q}(\phi^+ u \cdot n^+ + \phi^- u \cdot n^-)$ over interior facets. We could again use a conditional in order to represent the upwind value $\widetilde{q}$ by the correct choice of $q^+$ or $q^-$, depending on the sign of $u \cdot n^+$, say. Instead, we make use of the quantity un, which is either $u \cdot n$ or zero, in order to avoid writing explicit conditionals. Although it is not obvious at first sight, the expression given in code is equivalent to the desired expression, assuming $n^- = -n^+$.

```
L1 = dtc*(
    q*div(phi*u)*dx
    - conditional(dot(u, n) < 0, phi*dot(u, n)*q_in, 0.0)*ds
    - conditional(dot(u, n) > 0, phi*dot(u, n)*q, 0.0)*ds
    - (phi('+')-phi('-'))*(un('+')*q('+')-un('-')*q('-'))*dS
)
```

In our Runge-Kutta scheme, the first step uses $q^n$ to obtain $q^{(1)}$. We therefore declare similar forms that use $q^{(1)}$ to obtain $q^{(2)}$, and $q^{(2)}$ to obtain $q^{n+1}$. We make use of UFL's `replace` feature to avoid writing out the form repeatedly.

```
q1 = Function(V)
q2 = Function(V)
L2 = replace(L1, {q: q1})
L3 = replace(L1, {q: q2})
```

We now declare a variable to hold the temporary increments at each stage.

```
dq = Function(V)
```

Since we want to perform hundreds of time steps, ideally we should avoid reassembling the left-hand-side mass matrix each step, as this does not change. We therefore make use of the `LinearVariationalSolver` object for each of our Runge-Kutta stages. These cache and reuse the assembled left-hand-side matrix. Since the DG mass matrices are block-diagonal, we use an incomplete ILU factorisation as a preconditioner to solve the linear systems. As a minor technical point, we in fact use an outer block Jacobi preconditioner. This allows the code to be executed in parallel without any further changes being necessary.

```
params = {'ksp_type': 'preonly',
          'pc_type': 'bjacobi',
          'sub_pc_type': 'ilu'}
prb1 = LinearVariationalProblem(a, L1, dq)
solv1 = LinearVariationalSolver(prb1, solver_parameters=params)
prb2 = LinearVariationalProblem(a, L2, dq)
solv2 = LinearVariationalSolver(prb2, solver_parameters=params)
prb3 = LinearVariationalProblem(a, L3, dq)
solv3 = LinearVariationalSolver(prb3, solver_parameters=params)
```

We now run the time loop. This consists of three Runge-Kutta stages, and every 20 steps we write out the solution to file and print the current time to the terminal.

```
t = 0.0
step = 0
while t < T - 0.5*dt:
    solv1.solve()
    q1.assign(q + dq)

    solv2.solve()
    q2.assign(0.75*q + 0.25*(q1 + dq))

    solv3.solve()
    q.assign((1.0/3.0)*q + (2.0/3.0)*(q2 + dq))

    step += 1
    t += dt
```

```python

    if step % 20 == 0:
        outfile.write(q)
        print("t=", t)
```

Finally, we display the normalised $L^2$ error, by comparing to the initial condition. Figure 4.4 displays snapshots of the scalar field at select time-steps, for a complete period $T = 2\pi$.

```python
L2_err = errornorm(q, q_init, norm_type="L2")
L2_init = norm(q_init, norm_type="L2")
print(L2_err/L2_init)
```

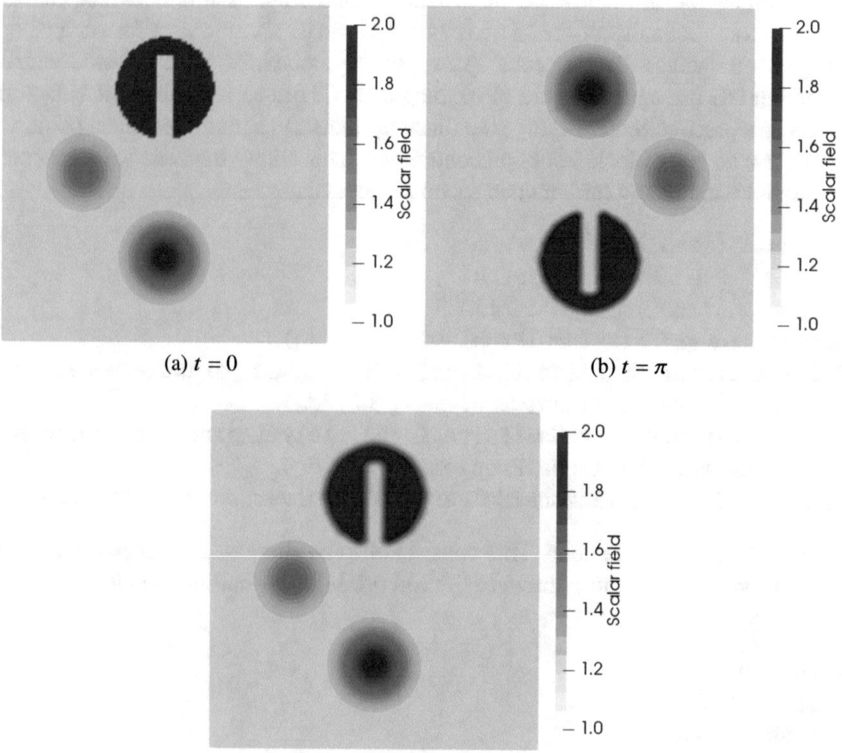


(a) $t = 0$ (b) $t = \pi$

(c) $t = 2\pi$

**Fig. 4.4** Evolution of the cosine-bell-cone-slotted-cylinder test. From (a)–(c): the advected scalar field at $t = 0$, $\pi$, and $2\pi$

## 4.3 Nonlinear Shallow Water Equations

We now introduce a nonlinear solver for the full shallow water system, using techniques outlined previously in Chapter 2. The main objective of this section is to introduce the reader to common solution schemes employed by modern dynamical cores for atmospheric and oceanic flows. Our approach revolves around using a quasi-Newton/Picard iteration method for computing linearised updates of the prognostic fields. In Section 2.3.4, we derived a compatible finite element scheme for the full nonlinear shallow water equations, with an implicit midpoint timestepping scheme.

We summarise the scheme as follows. Given a velocity field $u^n \in V_1$ and depth perturbation field $h^n \in V_2$ at time $t_n$, we define a sequence of approximations $v^i$, and $p^i$ to $u^{n+1}$ and $h^{n+1}$ respectively, we define the increments for the $k$-th Picard iteration $\delta v^k$ and $\delta p^k$, with $v^{k+1} = v^k + \delta v^k, \quad p^{k+1} = p^k + \delta p^k$, which are chosen to satisfy the linear finite element problem

$$\int_\Omega w \cdot \delta v^k \, \mathrm{d}x + \frac{\Delta t}{2} \int_\Omega w \cdot f \left(\delta v^k\right)^\perp \mathrm{d}x \\ - \frac{g\Delta t}{2} \int_\Omega \delta p^k \nabla \cdot w \, \mathrm{d}x = -R_u[w; v^k, p^k], \tag{4.12}$$

$$\frac{H\Delta t}{2} \int_\Omega \phi \nabla \cdot \delta v^k \, \mathrm{d}x + \int_\Omega \phi \, \delta p^k = -R_h[\phi; v^k, p^k]. \tag{4.13}$$

The residual functionals $R_u$ and $R_h$ are are constructed from the implicit midpoint discretisation, presented in equations 2.63–2.64.

To extend the stable timestep of the scheme, we make one further modification. We now replace $R_u$ and $R_h$ by equivalent functionals (i.e. functionals that also vanish when the implicit midpoint rule equations are satisfied). To do this, we set $\bar{u}^k = \frac{1}{2}(u^n + v^k)$ and $\bar{h}^k = \frac{1}{2}(u^n + h^k)$. Then we solve the implicit equations for $u^{n+1}$ and $h^{n+1}$ using $\bar{u}^k$ as the advecting velocity, *i.e.*, solve for $\hat{v}^{k+1}$, $\hat{p}^{k+1}$ such that

$$\int_\Omega w \cdot \frac{\hat{v}^{k+1} - u^n}{\Delta t} \, \mathrm{d}x + \int_\Omega w \cdot f(\hat{v}^{k+\frac{1}{2}})^\perp \, \mathrm{d}x \\ - \int_\Omega \nabla^\perp \left(w \cdot (\bar{u}^\perp)^k\right) \cdot \hat{v}^{k+\frac{1}{2}} \, \mathrm{d}x \\ + \int_\Gamma [n^\perp \left(w \cdot (\bar{u}^\perp)^k\right)] \cdot \widetilde{\hat{v}^{k+\frac{1}{2}}} \, \mathrm{d}s \\ + \int_\Omega \nabla \cdot w \left( g\left(\bar{h}^k + b\right) + \frac{1}{2}|\bar{u}^k|^2 \right) \mathrm{d}x = 0, \quad \forall w \in V_1, \tag{4.14}$$

and

$$\int_\Omega \phi \frac{\hat{p}^{k+1} - h^n}{\Delta t} \, \mathrm{d}x - \int_\Omega \nabla\phi \cdot \left( \bar{u}^k \hat{p}^{k+\frac{1}{2}} \right) \mathrm{d}x + \int_\Gamma [\![ \phi \bar{u}^k ]\!] \tilde{\hat{p}}^{k+\frac{1}{2}} \, \mathrm{d}s = 0, \quad \forall \phi \in V_2, \tag{4.15}$$

where $\hat{v}^{k+\frac{1}{2}} = \frac{1}{2}(\hat{v}^{k+1} + u^n)$, $\hat{h}^{k+\frac{1}{2}} = \frac{1}{2}(\hat{h}^{k+1} + h^n)$. This can be thought of as iteratively solving for the average velocity and depth that satisfies the implicit midpoint rule discretisation. The equations (4.14) and (4.15) can be solved separately, since there is no coupling between them. Finally, we define the new residuals

$$\hat{R}_u[w; v^k, p^k] = \int_\Omega w \cdot (v^k - \hat{v}^{k+1}) \, \mathrm{d}x, \tag{4.16}$$

$$\hat{R}_h[\phi; v^k, p^k] = \int_\Omega \phi (p^k - \hat{p}^{k+1}) \, \mathrm{d}x, \tag{4.17}$$

which replace $R_u$ and $R_h$ in (4.12) and (4.13). The complete nonlinear procedure is summarised in Algorithm 1.

---

**Algorithm 1** Nonlinear procedure for the rotating shallow water system using a semi-implicit time-integration scheme.

$t_n = 0$
$u^n \leftarrow u(x, t = 0)$ ▷ Initial condition for the velocity field
$h^n \leftarrow h(x, t = 0)$ ▷ Initial condition for the depth field
**while** $t < t_{\max}$ **do**
  $v^0 = u^n$ ▷ Initialise the Picard method
  $p^0 = h^n$
  **for** $k \in \{0, \cdots, k_{\max} - 1\}$ **do** ▷ Start Picard cycle
    Solve (4.15) and (4.14) for candidates $\hat{p}^k$ and $\hat{v}^k$.
    Use candidates to determine $\hat{R}_u^k$ and $\hat{R}_h^k$.
    Solve the discrete linear system:

$$\begin{bmatrix} M_{1,f} & -\frac{g\Delta t}{2} D^T \\ \frac{H\Delta t}{2} D & M_2 \end{bmatrix} \begin{Bmatrix} \delta v^k \\ \delta p^k \end{Bmatrix} = - \begin{Bmatrix} \hat{R}_u^k \\ \hat{R}_h^k \end{Bmatrix}$$

    $v^{k+1} \leftarrow v^k + \delta v^k$ ▷ Linear update for velocity
    $p^{k+1} \leftarrow p^k + \delta p^k$ ▷ Linear update for depth
  **end for**
  $t_n \leftarrow t_n + \Delta t$
  $u^n \leftarrow v^k$ ▷ Update velocity for next time step
  $h^n \leftarrow p^k$ ▷ Update depth for next time step
**end while**

---

We shall proceed to use this iterative approach to solve the nonlinear rotating shallow water equations for a more complicated test case. The test case simulating atmospheric flow over an isolated mountain (test case 5) is featured in Williamson et al. (1992). This examples does not have an analytic solution. It is, however, well studied by the geophysical fluid dynamics community.

### 4.3.1 Example: Nonlinear Flow Over an Isolated Mountain

The example starts with fields in geostrophic balance, as in the solid body rotation test (4.6). One difference here is that our maximal zonal winds are set to be $20\text{ms}^{-1}$. Moreover, we now introduce topography in the system describing an isolated mountain with its center at latitude $\phi_c = \pi/6$ and longitude $\lambda_c = -\pi/2$. The topography is given by the expression

$$b = 2000\left(1 - \frac{\left(\min\{R_l^2, (\phi-\phi_c)^2 + (\lambda-\lambda_c)^2\}\right)^{1/2}}{R_l}\right), \tag{4.18}$$

where $R_l = \pi/9$.

As the zonal flow interacts with the mountain, it produces waves that travel around the globe. As there is no analytical solution for this problem, obtaining a reference solution from an accepted-good model is necessary to study the convergence of the model. Such a study is conducted in Shipton et a. (2018), using a different formulation but a similar Picard method/semi-implicit approach as the one outlined in Algorithm 1.

Implementing this example in Firedrake follows similarly to the solid body rotation simulation in Section 4.1.1. In fact, the code is identical up until specifying the initial conditions (shown here for completeness):

```
from firedrake import *
R0 = 6371220.0
R = Constant(R0)                # Radius of the earth [m]
H = Constant(5960.0)            # Mean depth [m]
day = 24. * 60. * 60.           # Time in a day [s]
Omega = Constant(7.292E-5)      # Angular rotation rate [rads]
g = Constant(9.80616)           # Accel. due to gravity [m/s^s]
mesh_degree = 3     # Cubic coordinate field
refinements = 5     # Number of refinements
mesh = IcosahedralSphereMesh(R0, refinements,
                             degree=mesh_degree)
x = SpatialCoordinate(mesh)
global_normal = as_vector([x[0], x[1], x[2]])
mesh.init_cell_orientations(global_normal)
model_degree = 2    # Degree of the finite element complex
V1 = FunctionSpace(mesh, "BDM", model_degree)
V2 = FunctionSpace(mesh, "DG", model_degree - 1)
V = V1 * V2    # Mixed space for velocity and depth
f_expr = 2 * Omega * x[2] / R
Vf = FunctionSpace(mesh, "CG", mesh_degree)
f = Function(Vf).interpolate(f_expr)    # Coriolis frequency
```

The fields are initialised using UFL expressions as in the solid body rotation example, using the same set of physical constants, function spaces, mesh, and coordinate field. The only difference is in the prescribed mean zonal wind, which is $20\,\mathrm{ms}^{-1}$ for this example.

```
u_0 = 20.0  # maximum amplitude of the zonal wind [m/s]
u_max = Constant(u_0)
u_exp = as_vector([-u_max*x[1]/R, u_max*x[0]/R, 0.0])
h_exp = H-((R*Omega*u_max+u_max*u_max/2.0)*(x[2]*x[2]/(R*R)))/g
un = Function(V1).project(u_exp)
hn = Function(V2).interpolate(h_exp)
```

Setting up the topography is accomplished similarly. We use the UFL operator `Min` to construct expression in (4.18). We also convert to Cartesian coordinates $(x,y,z)$ from Earth-centered latitude-longitude coordinates $(\phi,\lambda)$, employing the `asin` and `atan_2` UFL built-in functions. We offset the initial depth field by the topography $b$ to get the depth field described over the surface.

```
# Topography
rl = pi/9.0
lambda_x = atan_2(x[1]/R0, x[0]/R0)
lambda_c = -pi/2.0
phi_x = asin(x[2]/R0)
phi_c = pi/6.0
minarg = Min(pow(rl, 2),
             pow(phi_x - phi_c, 2) +
             pow(lambda_x - lambda_c, 2))
bexpr = 2000.0*(1 - sqrt(minarg)/rl)
b = Function(V2, name="Topography").interpolate(bexpr)
hn -= b
```

Optionally, we can write out the topography to a file as we would any normal field. See Figure 4.5 for the resulting field on our triangulated mesh.

We run this simulation for a total of 15 days, though the flow becomes more turbulent past day 30. This example could be run up to day 50 with no difficulty. We prescribe a time step size of $\Delta = 45$ seconds. We also initialise `Function` objects for storing our nonlinear updates.

```
tmax = 15 * day       # Run for 15 days
Dt = 450.0
dt = Constant(Dt)     # Time-step size [s]
up = Function(V1)
hp = Function(V2)
```

Now we set up the linear variational solver for the depth equation (4.15) to determine the candidate depth field $\hat{h}$. We use direct solver configurations via an LU factorisation. We rewrite the jump term containing the upwind value of $h$ using a velocity expression which takes on the value of $u^{n+\frac{1}{2}} \cdot n$ where appropriate.

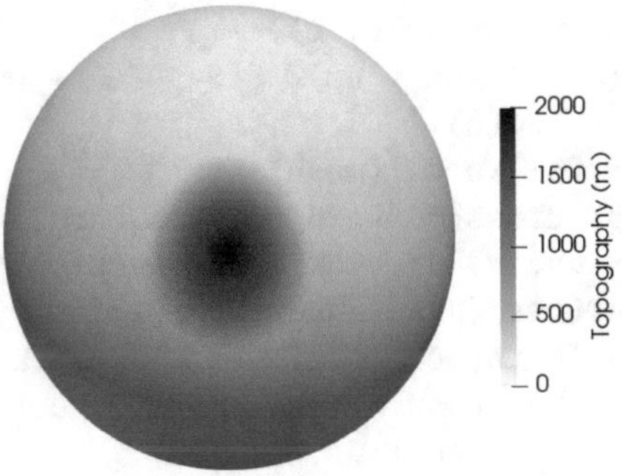


**Fig. 4.5** The topography field in meters

```
hps = Function(V2)
h = TrialFunction(V2)
phi = TestFunction(V2)
hh = 0.5 * (hn + h)
uh = 0.5 * (un + up)
n = FacetNormal(mesh)
uup = 0.5 * (dot(uh, n) + abs(dot(uh, n)))
Heqn = ((h - hn)*phi*dx - dt*inner(grad(phi), uh*hh)*dx
        + dt*jump(phi)*(uup('+')*hh('+')-uup('-')*hh('-'))*dS)
Hproblem = LinearVariationalProblem(lhs(Heqn), rhs(Heqn), hps)
lu_params = {'ksp_type': 'preonly',
             'pc_type': 'lu',
             'pc_factor_mat_solver_type': 'mumps'}
Hsolver = LinearVariationalSolver(Hproblem,
                                  solver_parameters=lu_params,
                                  options_prefix="H-advection")
```

The velocity equation in (4.14) is incorporated as a separate solver for the candidate $\hat{u}$. While the integral forms are more involved, we can use existing UFL operators to incorporate the surface term. For example, we create the operator $[v] = v^+ + v^-$ by using a custom function `both`, using the `avg` UFL operator. This allows us to include both the positive (+) and negative (-) restrictions in the integrand. We add in the Coriolis term using the same `perp` operator from the solid body rotation example. The solver is configured to use a direct LU factorisation.

```
ups = Function(V1)
u = TrialFunction(V1)
v = TestFunction(V1)
hh = 0.5 * (hn + hp)
ubar = 0.5 * (un + up)
uup = 0.5 * (dot(ubar, n) + abs(dot(ubar, n)))
uh = 0.5 * (un + u)
```

```
Upwind = 0.5 * (sign(dot(ubar, n)) + 1)
K = 0.5 * (inner(0.5 * (un + up), 0.5 * (un + up)))
both = lambda u: 2*avg(u)
outward_normals = CellNormal(mesh)
perp = lambda arg: cross(outward_normals, arg)
Ueqn = (inner(u - un, v)*dx + dt*inner(perp(uh)*f, v)*dx
        - dt*inner(perp(grad(inner(v, perp(ubar)))), uh)*dx
        + dt*inner(both(perp(n)*inner(v, perp(ubar))),
                   both(Upwind*uh))*dS
        - dt*div(v)*(g*(hh + b) + K)*dx)
Uproblem = LinearVariationalProblem(lhs(Ueqn), rhs(Ueqn), ups)
Usolver = LinearVariationalSolver(Uproblem,
                                  solver_parameters=lu_params,
                                  options_prefix="U-advection")
```

With linear solvers created for advecting the velocity and depth equations, we now set up the implicit linear solver for the updates $\delta u^k$ and $\delta h^k$. This will be solved repeated during the Picard cycle, for each time step. The bilinear form defining the linearised matrix operator is identical to that of the solid body rotation example in (4.5). The key difference lies in how the residuals are constructed, see equations (4.16) and (4.17). Again, we use a direct LU solver to invert the mixed system.

```
HU = Function(V)
deltaU, deltaH = HU.split()
w, phi = TestFunctions(V)
du, dh = TrialFunctions(V)
alpha = 0.5
HUlhs = (inner(w, du + alpha*dt*f*perp(du))*dx
         - alpha*dt*div(w)*g*dh*dx
         + phi*(dh + alpha*dt*H*div(du))*dx)
HUrhs = -inner(w, up - ups)*dx - phi*(hp - hps)*dx
HUproblem = LinearVariationalProblem(HUlhs, HUrhs, HU)
params = {'ksp_type': 'preonly',
          'mat_type': 'aij',
          'pc_type': 'lu',
          'pc_factor_mat_solver_type': 'mumps'}
HUsolver = LinearVariationalSolver(HUproblem,
                                   solver_parameters=params,
                                   options_prefix="impl-solve")
```

Before we start the time-integration loop, let us write out the initial fields for visualisation. Note that we add in the topography to the depth field to get the total surface elevation.

```
u_out = Function(V1, name="Velocity").assign(un)
h_out = Function(V2, name="Surface elevation").assign(hn + b)
outfile = File("results/W5/nonlinear_w5.pvd")
outfile.write(u_out, h_out)
```

Finally, we are ready to write the time-integration loop. The code below is an implementation of Algorithm 1; in each time step we compute the candidate fields by taking an advection step, followed by a sequence of updates via the solution of the linearised system. In our example, we set the maximum number of Picard iterations to be 4.

```
t = 0.0
dumpfreq = 10    # Dump output every 10 steps
counter = 1
k_max = 4        # Maximum number of Picard iterations
while t < tmax:
    t += Dt
    up.assign(un)
    hp.assign(hn)

    # Start picard cycle
    for i in range(k_max):
        # Advect to get candidates
        Hsolver.solve()
        Usolver.solve()

        # Linear solve for updates
        HUsolver.solve()

        # Increment updates
        up += deltaU
        hp += deltaH

    # Update fields for next time step
    un.assign(up)
    hn.assign(hp)

    counter += 1
    if counter > dumpfreq:
        u_out.assign(un)
        h_out.assign(hn + b)
        outfile.write(u_out, h_out)
        counter -= dumpfreq
```

We can view the output files as before. We have only run the simulation for 15 days, which is long enough to observe the nonlinearity of the flow. As the fluid passes over the mountain, compounded by the effects of rotation, waves are generated which distorts the fluid depth. Figure 4.6 displays snapshots of the depth of the fluid and velocity for day 5, 10, and 15.

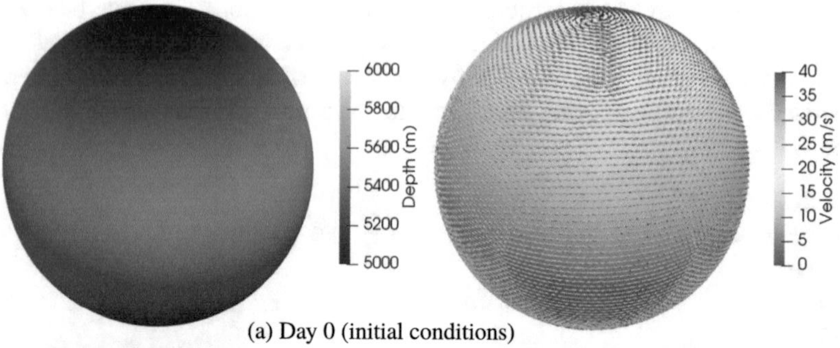


(a) Day 0 (initial conditions)

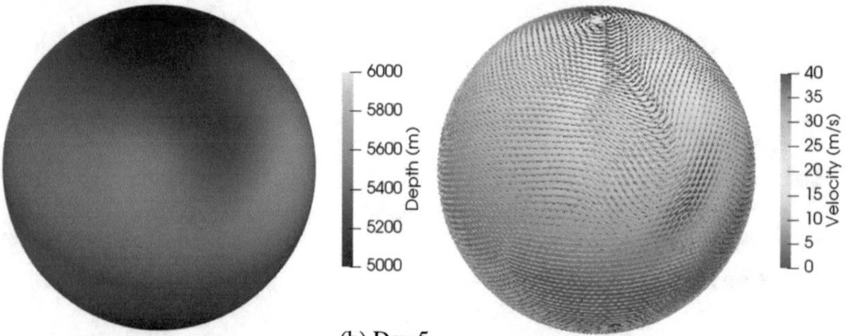


(b) Day 5

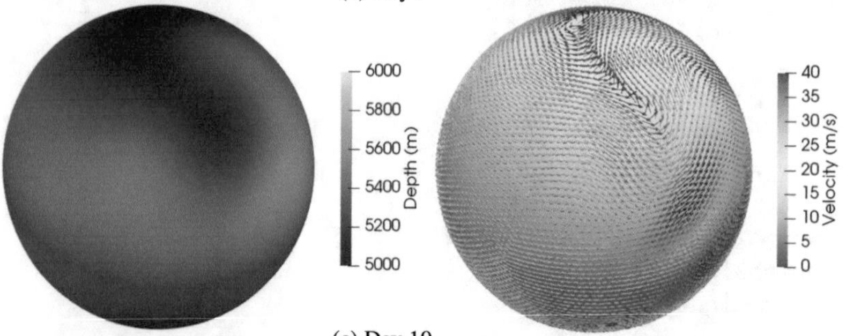


(c) Day 10

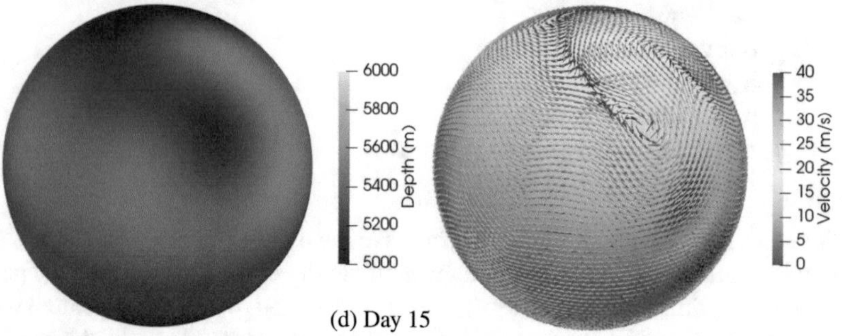


(d) Day 15

**Fig. 4.6** Snapshots from the isolated mountain (same orientation as in Figure 4.5) test case. The depth and velocity fields at day 0, 5, 10, and 15

# Chapter 5
# Models in Three-Dimensions

As we move from two dimensional models, such as the shallow water equations discussed in the last chapter, to three dimensional models, the problem of efficient parallel iterative algorithms for implicit linear systems becomes an urgent priority. In this chapter we discuss some situations where these algorithms are required in the context of compatible finite element methods applied to three dimensional models, and explain how they can be treated in Firedrake.

## 5.1 A Linear Compressible Model

In this section we consider the compressible Boussinesq equations, which exhibit non-hydrostatic and compressible effects whilst avoiding the full complexity of the compressible Euler equations pressure gradient term and equation of state. This model was used by Skamarock and Klemp (1994), for example, to explore aspects of time integration in a simplified setting.

The full equations are

$$\left(\frac{\partial}{\partial t} + \boldsymbol{u} \cdot \nabla\right) \boldsymbol{u} + f \hat{z} \times \boldsymbol{u} = \nabla p + b \hat{z}, \tag{5.1}$$

$$\left(\frac{\partial}{\partial t} + \boldsymbol{u} \cdot \nabla\right) p = -c^2 \nabla \cdot \boldsymbol{u}, \tag{5.2}$$

$$\left(\frac{\partial}{\partial t} + \boldsymbol{u} \cdot \nabla\right) b = 0, \tag{5.3}$$

where $\boldsymbol{u}$ is the velocity, $p$ is the pressure, $b$ is the buoyancy (how light a parcel of fluid is compared to a reference value), $f$ is the Coriolis parameter, $\hat{z}$ is the upward normal, and $c$ the speed of sound. In the limit $c \to \infty$, we recover the incompressible Boussinesq equations, which neglect variations in density in every single term except for the gravitational force.



T. H. Gibson et al., *Compatible Finite Element Methods for Geophysical Flows*, Mathematics of Planet Earth, https://doi.org/10.1007/978-3-030-23957-2_5

As we saw in the previous chapter, a semi-implicit formulation of a geophysical fluid model can be built up from an implicit midpoint rule discretisation of the equations linearised about a state of rest. Hence, we consider the following equations,

$$\frac{\partial \boldsymbol{u}}{\partial t} = \nabla p + b\hat{\boldsymbol{z}}, \tag{5.4}$$

$$\frac{\partial p}{\partial t} = -c^2 \nabla \cdot \boldsymbol{u}, \tag{5.5}$$

$$\frac{\partial b}{\partial t} = -N^2 \boldsymbol{u} \cdot \hat{\boldsymbol{z}}, \tag{5.6}$$

where $N$ the buoyancy frequency, which we call linear gravity wave system. For simplicity we assume that both $c$ and $N$ are constant, and enforce the slip boundary condition,

$$\boldsymbol{u} \cdot \boldsymbol{n} = 0, \tag{5.7}$$

at the upper and lower boundary of the atmosphere ($\partial\Omega_{\text{top}}$, $\partial\Omega_{\text{bottom}}$ respectively). This linearised system was summarised in Section 2.4, but we shall derive the full discrete system here for context.

### 5.1.1 Compatible Finite Element Formulation

The finite element discretisation of these equations begins with the specification of the finite element spaces to which the three variables $\boldsymbol{u}$, $b$ and $p$ are restricted. We construct the relevant finite element spaces from the following de-Rham complexes in one, two and three dimensions:

$$V_0 \xrightarrow{\partial_z} V_1, \qquad U_0 \xrightarrow{\nabla^\perp} U_1 \xrightarrow{\nabla\cdot} U_2, \qquad W_0 \xrightarrow{\nabla} W_1 \xrightarrow{\nabla\times} W_2 \xrightarrow{\nabla\cdot} W_3, \tag{5.8}$$

where

$$\begin{aligned} W_0 &= U_0 \otimes V_0, \\ W_1 &= \texttt{HCurl}(U_1 \otimes V_0) \oplus \texttt{HCurl}(U_0 \otimes V_1), \\ W_2 &= \texttt{HDiv}(U_2 \otimes V_0) \oplus \texttt{HDiv}(U_1 \otimes V_1), \\ W_3 &= U_2 \otimes V_1. \end{aligned} \tag{5.9}$$

The operations `HCurl` and `HDiv` are symbolic operators denoting how functions are pulled back to the reference element (see McRae et al. 2016 for implementation details). The space $W_2^0$ denotes the space of all functions in $W_2$ which satisfy the boundary condition (5.7). We also define the horizontal and vertical spaces $W_2^h \equiv \texttt{HDiv}(U_1 \otimes V_1)$ and $W_2^v \equiv \texttt{HDiv}(U_2 \otimes V_0)$. With these definitions, we observe the following decomposition: $W_2 = W_2^h \oplus W_2^v$. We seek the unknown fields from (5.4)–(5.6) in the following finite element spaces:

$$\boldsymbol{u} \in W_2^0 = W_2^h \oplus W_2^{v,0}, \qquad p \in W_3, \qquad b \in W_b, \tag{5.10}$$

where $W_b \equiv U_2 \otimes V_0$, and $W_2^{v,0}$ is the vertical component of $W_2$ with velocities vanishing on $\Omega_{\text{top}}$ and $\Omega_{\text{bottom}}$. Note that $W_2^v = \texttt{HDiv}(W_b)$. That is, $W_b$ and $W_2^v$ have the same number of degrees of freedom, but differ in how they are pulled back to the reference element. For ease of notation, we write $W_2$ in place of $W_2^0$. Similarly for the space $W_2^{v,0}$.

We remark here that the choice for $W_b$ in (5.10) corresponds to a Charney-Phillips vertical staggering of the buoyancy variable. One could also collocate $b$ with $p$ ($b \in W_3$), which corresponds to a Lorenz staggering. This however supports a computational mode which is exacerbated by fast-moving waves. We restrict our discussion to the former case.

To obtain the discrete system, we simply multiply equations (5.4)–(5.6) by test functions $\boldsymbol{w} \in W_2$, $\phi \in W_3$ and $\gamma \in W_b$ and integrate by parts. Using the Crank-Nicholson method (equivalent to the implicit midpoint rule in this case) to discretise in time, we obtain:

$$\int_\Omega \boldsymbol{w} \cdot \frac{\boldsymbol{u}^{(n+1)} - \boldsymbol{u}^{(n)}}{\Delta t} \,\mathrm{d}\boldsymbol{x} - \int_\Omega \nabla \cdot \boldsymbol{w} \frac{p^{(n+1)} + p^{(n)}}{2} \,\mathrm{d}\boldsymbol{x} - \int_\Omega \boldsymbol{w} \cdot \frac{b^{(n+1)} + b^{(n)}}{2} \hat{\boldsymbol{z}} \,\mathrm{d}\boldsymbol{x} = 0 \tag{5.11}$$

$$\int_\Omega \phi \frac{p^{(n+1)} - p^{(n)}}{\Delta t} \,\mathrm{d}\boldsymbol{x} + c^2 \int_\Omega \phi, \nabla \cdot \frac{\boldsymbol{u}^{(n+1)} + \boldsymbol{u}^{(n)}}{2} \,\mathrm{d}\boldsymbol{x} = 0 \tag{5.12}$$

$$\int_\Omega \gamma \frac{b^{(n+1)} - b^{(n)}}{\Delta t} \,\mathrm{d}\boldsymbol{x} + N^2 \int_\Omega \gamma \frac{\boldsymbol{u}^{(n+1)} + \boldsymbol{u}^{(n)}}{2} \cdot \hat{\boldsymbol{z}} \,\mathrm{d}\boldsymbol{x} = 0. \tag{5.13}$$

We introduce the increments $\delta \boldsymbol{u} \equiv \boldsymbol{u}^{(n+1)} - \boldsymbol{u}^{(n)}$, and set $\boldsymbol{u}_0 \equiv \boldsymbol{u}^{(n)}$ (similarly for $\delta p$, $p_0$, $\delta b$, and $b_0$).

In each time-step, we solve the following linear variational problem for the increments $\delta \boldsymbol{u}$, $\delta p$ and $\delta b$: find $\delta \boldsymbol{u} \in W_2$, $\delta p \in W_3$ and $\delta b \in W_b$ such that

$$\int_\Omega \boldsymbol{w} \cdot \delta \boldsymbol{u} \,\mathrm{d}\boldsymbol{x} - \frac{\Delta t}{2} \int_\Omega \nabla \cdot \boldsymbol{w} \delta p \,\mathrm{d}\boldsymbol{x} - \frac{\Delta t}{2} \int_\Omega \boldsymbol{w} \cdot \delta b \hat{\boldsymbol{z}} \,\mathrm{d}\boldsymbol{x} = R_u, \tag{5.14}$$

$$\int_\Omega \phi \delta p \,\mathrm{d}\boldsymbol{x} + \frac{\Delta t}{2} c^2 \int_\Omega \phi \nabla \cdot \delta \boldsymbol{u} \,\mathrm{d}\boldsymbol{x} = R_\phi, \tag{5.15}$$

$$\int_\Omega \gamma \delta b \,\mathrm{d}\boldsymbol{x} + \frac{\Delta t}{2} N^2 \int_\Omega \gamma \delta \boldsymbol{u} \cdot \hat{\boldsymbol{z}} \,\mathrm{d}\boldsymbol{x} = R_b, \tag{5.16}$$

for all $\boldsymbol{w} \in W_2$, $\phi \in W_3$ and $\gamma \in W_b$, where

$$R_u = \Delta t \int_\Omega \nabla \cdot \boldsymbol{w} p_0 \,\mathrm{d}\boldsymbol{x} + \Delta t \int_\Omega \boldsymbol{w} b_0 \hat{\boldsymbol{z}} \,\mathrm{d}\boldsymbol{x} \tag{5.17}$$

$$R_p = -\Delta t c^2 \int_\Omega \phi \nabla \cdot \boldsymbol{u}_0 \,\mathrm{d}\boldsymbol{x} \tag{5.18}$$

$$R_b = -\Delta t N^2 \int_\Omega \gamma \boldsymbol{u}_0 \cdot \hat{\boldsymbol{z}} \,\mathrm{d}\boldsymbol{x}. \tag{5.19}$$

The system of equations (5.14)–(5.16) can be written as a matrix equation for the degree of freedom (DOF) coefficient vectors $\boldsymbol{U}$ (velocity), $\boldsymbol{P}$ (pressure) and $\boldsymbol{B}$ (buoyancy):

$$\begin{bmatrix} M_{\boldsymbol{u}} & -\frac{\Delta t}{2}D^T & -\frac{\Delta t}{2}Q \\ \frac{\Delta t}{2}c^2 D & M_p & 0 \\ \frac{\Delta t}{2}N^2 Q^T & 0 & M_b \end{bmatrix} \begin{Bmatrix} \boldsymbol{U} \\ \boldsymbol{P} \\ \boldsymbol{B} \end{Bmatrix} = \begin{Bmatrix} \boldsymbol{R_u} \\ \boldsymbol{R}_p \\ \boldsymbol{R}_b \end{Bmatrix}, \tag{5.20}$$

where $\boldsymbol{R_u}, \boldsymbol{R}_p$, and $\boldsymbol{R}_b$ are the coefficient vectors for the right-hand side residuals. The rest of this section will focus on solution approaches for solving (5.20).

### 5.1.2 Implicit Solver Strategy

In this section, we describe the solver strategy for the discrete linear gravity wave system 5.20. The main challenge is that it is a block-matrix system, so LU factorisation leads to a lot of "fill-in"; the number of non-sparse entries in L and U increases in proportion to the number of cells squared. This means that direct solution rapidly becomes unviable with increasing resolution.

This means that we must turn to iterative solvers, in the form of preconditioned Krylov subspace methods. Again, the block-matrix structure means that standard off-the-shelf preconditioners that work with arbitrary sparse matrices (Jacobi or Gauss-Seidel iteration, for example) are not efficient preconditioners for our system. We will address this in this section by discussing elimination methods that reduce the system to a single $1 \times 1$ block that can be efficiently preconditioned by standard methods.

We begin by eliminating the buoyancy variable. From (5.20), we can use the third equation for buoyancy to arrive at an elimination process, which writes $\boldsymbol{B}$ as a function of $\boldsymbol{U}$:

$$M_b \boldsymbol{B} + \frac{\Delta t}{2} N^2 Q^T \boldsymbol{U} = \boldsymbol{R}_b \quad \Longrightarrow \quad \boldsymbol{B} = M_b^{-1} \boldsymbol{R}_b - \frac{\Delta t}{2} N^2 M_b^{-1} Q^T \boldsymbol{U}. \tag{5.21}$$

In PDE form, this is simply stating that the point-wise relation:

$$\delta b + \frac{\Delta t}{2} N^2 \delta \boldsymbol{u} \cdot \hat{\boldsymbol{z}} = r_b \tag{5.22}$$

holds. Substituting (5.21) into the discrete system (5.20), we obtain a mixed set of equations for the velocity and pressure variables:

$$\boldsymbol{A} \begin{Bmatrix} \boldsymbol{U} \\ \boldsymbol{P} \end{Bmatrix} = \begin{bmatrix} \widetilde{M}_{\boldsymbol{u}} & -\frac{\Delta t}{2}D^T \\ c^2 \frac{\Delta t}{2} D & M_p \end{bmatrix} \begin{Bmatrix} \boldsymbol{U} \\ \boldsymbol{P} \end{Bmatrix} = \begin{Bmatrix} \widetilde{\boldsymbol{R}}_{\boldsymbol{u}} \\ \boldsymbol{R}_p \end{Bmatrix}, \tag{5.23}$$

where

$$\widetilde{M}_{\boldsymbol{u}} = M_{\boldsymbol{u}} + \frac{\Delta t^2}{4} N^2 Q M_b^{-1} Q^T \tag{5.24}$$

is the modified mass matrix for the velocity, and the new right-hand side:

$$\widetilde{\boldsymbol{R}}_{\boldsymbol{u}} = \boldsymbol{R}_{\boldsymbol{u}} + \frac{\Delta t}{2} Q M_b^{-1} \boldsymbol{R}_b. \tag{5.25}$$

In the absence of orography, the operator relation holds: $Q^T = M_b \Pi_v$, where $\Pi_v : \mathbb{W}_2 \rightarrow \mathbb{W}_2^v$ is the projection onto the vertical component of the $\mathbb{W}_2$ space. With this in mind, the modified velocity mass matrix in (5.24) becomes:

$$\widetilde{M}_{\boldsymbol{u}} = M_{\boldsymbol{u}} + \frac{\Delta t^2}{4} N^2 \Pi_v^T M_b^{-1} \Pi_v. \tag{5.26}$$

This conveniently corresponds to the integral expression:

$$\{\Pi_v^T M_b^{-1} \Pi_v\}_{ij} = \int_\Omega \boldsymbol{w}_i \cdot \hat{z} \boldsymbol{w}_j \cdot \hat{z} \, \mathrm{d}\boldsymbol{x}. \tag{5.27}$$

The advantage here is that this does not require the inversion of $M_b$ when constructing the modified mass operator, making this approach much cheaper. That such an operation makes sense stems from the fact that the number of degrees of freedom in $\mathbb{W}_b$ is equal to that of $\mathbb{W}_2^v$ by construction.

We remark that this elimination strategy works without orography because (5.22) is satisfied in a point-wise sense. This does not hold with orography and therefore requires computing $M_b^{-1}$ when projecting out the buoyancy. Fortunately, $M_b$ is well-conditioned and can be inverted with a small number of Richardson or conjugate gradient iterations.

The solver strategy for solving the full gravity wave system can be summarised in two-stages. During each time-step, we perform the following:

1. Solve (5.23) for $\boldsymbol{U}$ and $\boldsymbol{P}$.
2. Using the computed velocity field, reconstruct $\boldsymbol{B}$ by solving the mass matrix system in (5.21).

Reconstructing the buoyancy is straight-forward; since this only involves inverting a mass matrix, the conjugate gradient algorithm with modest preconditioning will suffice. Inverting (5.20) is more troublesome. The mixed velocity-pressure system forms a saddle-point set of equations, which requires extensive preconditioning for a nonsymmetric Krylov method (such as GMRES) to converge.

## 5.2 Preconditioning Mixed Finite Element Systems

The primary solver strategy here is to use an "outer" Krylov method to iteratively solve (5.20) to a prescribed tolerance. In most operational weather forecasting codes, this system is solved to about four or five digits of accuracy ($r_{\text{tol}} = 10^{-4}$, for example). That is, we compute an approximation to the state $\boldsymbol{x}^k = \left\{\boldsymbol{U}^k \; \boldsymbol{P}^k\right\}^T$ such that

$$\frac{\| \boldsymbol{R} - \boldsymbol{A}\boldsymbol{x}^k \|_2}{\| \boldsymbol{R} \|_2} \leq r_{\text{tol}}, \tag{5.28}$$

where $\boldsymbol{A}$ is the left-hand side *mixed operator* in (5.20), $\boldsymbol{R} = \left\{\widetilde{\boldsymbol{R}}_{\boldsymbol{u}}\ \boldsymbol{R}_p\right\}^T$, and $\| \cdot \|_2$ is the standard vector 2-norm. Equation (5.28) is the termination criteria for the outer Krylov method. Since $\boldsymbol{A}$ is an indefinite matrix, this requires Krylov subspace methods designed for general matrices, such as the generalised minimal residual method (GMRES).

In most serious applications, Krylov methods rarely converge without proper motivation from a good preconditioner. In a general sense, a *preconditioner* is a transformation (possibly nonlinear) applied to the residual of the linear system: $\boldsymbol{r}^k \equiv \boldsymbol{R} - \boldsymbol{A}\boldsymbol{x}^k$, where $\boldsymbol{x}^k$ is the approximation generated at the $k$-th Krylov iteration. Applying a preconditioner, $\mathcal{P}$, we solve the equivalent problem:

$$\mathcal{P}\boldsymbol{r}^k = \mathcal{P}\left(\boldsymbol{R} - \boldsymbol{A}\boldsymbol{x}^k\right) = 0, \tag{5.29}$$

which produces an error $\epsilon^k \equiv \boldsymbol{x} - \boldsymbol{x}^k$ at the $k$-th iteration. If $\mathcal{P}$ is an application of a direct inverse, such as an LU or QR factorisation method, then $\epsilon^k$ is an exact error (up to numerical round-off) and the outer Krylov method will converge in a single iteration.

Direct inversion of $\boldsymbol{A}$ is impractical for most applications. We can, however, use basic matrix factorisations to build an *approximate* inverse to accelerate the convergence of, say, GMRES on the outer system.

### 5.2.1 Schur Complement Preconditioner

Preconditioners based on the Schur complement factorisation of $\boldsymbol{A}$ in (5.23) can be quite robust (Benzi et al., 2005). We start with the inverse of the Schur complement factorisation of $\boldsymbol{A}$ (in exact arithmetic):

$$\boldsymbol{A}^{-1} = \begin{bmatrix} I & \frac{\Delta t}{2}\widetilde{M}_{\boldsymbol{u}}^{-1}D^T \\ 0 & I \end{bmatrix} \begin{bmatrix} \widetilde{M}_{\boldsymbol{u}}^{-1} & 0 \\ 0 & H^{-1} \end{bmatrix} \begin{bmatrix} I & 0 \\ -c^2\frac{\Delta t}{2}D\widetilde{M}_{\boldsymbol{u}}^{-1} & I \end{bmatrix}, \tag{5.30}$$

where $H$ is the elliptic operator defined as: $H = M_p + c^2\frac{\Delta t^2}{4}D\widetilde{M}_{\boldsymbol{u}}^{-1}D^T$. The pressure is computed by solving the discrete "Helmholtz" problem:

$$H\boldsymbol{P} = \widetilde{\boldsymbol{R}}_p, \tag{5.31}$$

where

$$\widetilde{\boldsymbol{R}}_p = \boldsymbol{R}_p - c^2\frac{\Delta t}{2}D\widetilde{M}_{\boldsymbol{u}}^{-1}\widetilde{\boldsymbol{R}}_{\boldsymbol{u}}, \tag{5.32}$$

with $\widetilde{\boldsymbol{R}}_{\boldsymbol{u}}$ as defined in (5.25). The main difficulty here is that $H$ is dense since it contains the dense inverse $\widetilde{M}_{\boldsymbol{u}}^{-1}$. This is due to the fact that the normal components of basis functions in $\mathbb{W}_2$ are continuous between inter-elemental boundaries. Therefore, in practice, preconditioners designed around (5.30) rely on sufficient approximations to $\widetilde{M}_{\boldsymbol{u}}$ (hence $\widetilde{M}_{\boldsymbol{u}}^{-1}$) which renders $H$ sparse.

Rather than using the full matrix $\widetilde{M}_{\boldsymbol{u}}$, we can build a preconditioner using the inverse of a diagonalised velocity mass matrix:

$$\{\widetilde{\mathtt{M}}_{\boldsymbol{u}}^{-1}\}_{ij} = \frac{\delta_{ij}}{\{\widetilde{M}_{\boldsymbol{u}}\}_{ii}}. \tag{5.33}$$

We define our preconditioner via:

$$\mathcal{P} = \begin{bmatrix} I & \frac{\Delta t}{2}\widetilde{M}_{\boldsymbol{u}}^{-1}D^T \\ 0 & I \end{bmatrix} \begin{bmatrix} \widetilde{M}_{\boldsymbol{u}}^{-1} & 0 \\ 0 & \widetilde{H}^{-1} \end{bmatrix} \begin{bmatrix} I & 0 \\ -c^2\frac{\Delta t}{2}D\widetilde{M}_{\boldsymbol{u}}^{-1} & I \end{bmatrix}, \tag{5.34}$$

where

$$\widetilde{H} = M_p + c^2\frac{\Delta t^2}{4}D\widetilde{\mathtt{M}}_{\boldsymbol{u}}^{-1}D^T. \tag{5.35}$$

Since $\widetilde{M}_{\boldsymbol{u}}^{-1}$ is just a modified mass matrix, we can invert this approximately using bloc-ILU, for example. Similarly, $\widetilde{H}$ is now a *sparse* approximation to the symmetric elliptic operator $H$, hence we may employ standard solver strategies such as multigrid to invert the operator. We will demonstrate how one invokes such a preconditioner in our numerical example.

### 5.2.2 Example: Gravity Waves on a Small Planet

First, we import Firedrake and create some relevant physical constants.

```
from firedrake import *
nlayers = 20                    # Number of extrusion layers
R = 6.371E6/125.0               # Scaled radius [m]: R_earth/125.0
thickness = 1.0E4               # Thickness [m] of the domain
degree = 1                      # Degree of finite element complex
refinements = 4                 # Number of horizontal refinements
c = Constant(343.0)             # Speed of sound
N = Constant(0.01)              # Buoyancy frequency
Omega = Constant(7.292E-5)      # Angular rotation rate
dt = 36.0                       # Time-step size
tmax = 3600.0                   # End time
```

Then, we construct the mesh, which is obtained by taking a icosahedral meshing of a sphere with radius 125 times smaller than Earth, extruded upwards by $10^4$ m. This results in a mesh of 20 layers of triangular prism elements.

```
# Horizontal base mesh (cubic coordinate field)
base = IcosahedralSphereMesh(R,
```

```
                                    refinement_level=refinements,
                                    degree=3)

# Extruded mesh
mesh = ExtrudedMesh(base,
                    extrusion_type='radial',
                    layers=nlayers,
                    layer_height=thickness/nlayers)
```

We then build the necessary finite elements, which are constructed as tensor products of the following horizontal and vertical elements.

```
# Create tensor product complex:
# Horizontal elements
U1 = FiniteElement('RT', triangle, degree)
U2 = FiniteElement('DG', triangle, degree - 1)

# Vertical elements
V0 = FiniteElement('CG', interval, degree)
V1 = FiniteElement('DG', interval, degree - 1)
```

These are then combined using a symbolic algebra layer to form the tensor product elements for the velocity, pressure and buoyancy variables. More details on the mechanism behind these constructions can be found in McRae et al. (2016).

```
# HDiv element
W2_ele_h = HDiv(TensorProductElement(U1, V1))
W2_ele_v = HDiv(TensorProductElement(U2, V0))
W2_ele = W2_ele_h + W2_ele_v

# L2 element
W3_ele = TensorProductElement(U2, V1)

# Charney-Phillips element
Wb_ele = TensorProductElement(U2, V0)
```

The resulting tensor product finite element spaces are defined in the usual way using UFL:

```
# Resulting function spaces
W2 = FunctionSpace(mesh, W2_ele)
W3 = FunctionSpace(mesh, W3_ele)
Wb = FunctionSpace(mesh, Wb_ele)
```

Next we set up the initial conditions, for which we need the coordinate field.

```
x = SpatialCoordinate(mesh)
```

The initial condition for our velocity will be a simple solid body rotation field,

$$u = 20e_\lambda, \tag{5.36}$$

where $e_\lambda$ is the unit vector pointing in the direction of decreasing longitude. This is expressed as follows:

```
# Initial condition for velocity
u0 = Function(W2)
u_max = Constant(20.0)
uexpr = as_vector([-u_max*x[1]/R, u_max*x[0]/R, 0.0])
u0.project(uexpr)
```

The initial condition for the buoyancy is a localised anomaly given by

$$b = \frac{d^2}{d^2 + q^2}\sin(2\pi z/L_z), \quad q = R\cos^{-1}(\cos(\phi)\cos(\lambda - \lambda_c)), \tag{5.37}$$

where $d = 5000\,\mathrm{m}$, $L_z = 20000$, $R = 6371\,\mathrm{km}/125$ is the planet radius, and $\lambda_c = 2/3$. As in previous examples, we can construct (5.37) by composing various UFL expressions together:

```
# Initial condition for the buoyancy perturbation
lamda_c = 2.0*pi/3.0
phi_c = 0.0
W_CG1 = FunctionSpace(mesh, "CG", 1)
z = Function(W_CG1).interpolate(
    sqrt(x[0]*x[0] + x[1]*x[1] + x[2]*x[2]) - R
)
lat = Function(W_CG1).interpolate(
    asin(x[2]/sqrt(x[0]*x[0] + x[1]*x[1] + x[2]*x[2]))
)
lon = Function(W_CG1).interpolate(atan_2(x[1], x[0]))
b0 = Function(Wb)
L_z = 20000.0
d = 5000.0
sin_tmp = sin(lat) * sin(phi_c)
cos_tmp = cos(lat) * cos(phi_c)
q = R*acos(sin_tmp + cos_tmp*cos(lon-lamda_c))
s = (d**2)/(d**2 + q**2)
bexpr = s*sin(2*pi*z/L_z)
b0.interpolate(bexpr)
```

The pressure is just initialised to zero.

```
# Initial condition for pressure
p0 = Function(W3).assign(0.0)
```

Next we move on to formulating the finite element solvers. We build a mixed function space for $\boldsymbol{u}$ and $p$ and set up test and trial functions.

```
# Set up linear variational solver for u-p
# (After eliminating buoyancy)
W = W2 * W3
u, p = TrialFunctions(W)
w, phi = TestFunctions(W)
```

We also set up the radial unit vector used in the eliminated buoyancy term and reconstruction later on.

```
# radial unit vector
khat = interpolate(x/sqrt(dot(x, x)),
                   mesh.coordinates.function_space())
```

We express Equations (5.39–5.40) using the test and trial functions.

```
a_up = (dot(w, u)
        - 0.5*dt*p*div(w)
        # Appears after eliminating b
        + (0.5*dt*N)**2*dot(w, khat)*dot(u, khat)
        + phi*p + 0.5*dt*c**2*phi*div(u))*dx

L_up = (dot(w, u0)
        + 0.5*dt*dot(w, khat*b0)
        + phi*p0)*dx
```

Next, we handle the boundary conditions, which are slip conditions $\boldsymbol{u} \cdot \boldsymbol{n} = 0$ on the top and bottom boundaries of the atmosphere for $\boldsymbol{u}$. It may appear like we are setting both components of $\boldsymbol{u}$ but as we are using an H(div) element, only normal components of $\boldsymbol{u}$ are associated with boundary entities, so only these are set accordingly.

```
bcs = [DirichletBC(W.sub(0), 0.0, "bottom"),
       DirichletBC(W.sub(0), 0.0, "top")]
```

Now we set up a function to solve the mixed system into and define the symbolic problem.

```
w = Function(W)
up_problem = LinearVariationalProblem(a_up, L_up, w, bcs=bcs)
```

Here, we construct a solver objcet to be re-used every time-step. We now wish to utilise a specific set of configurations to solve our linear system. We can do this by specifying both our iterative solver mixed preconditioner directly to PETSc. Proceeding line-by-line:

```
103 params = {
104     'pc_type': 'fieldsplit',
105     'pc_fieldsplit_type': 'schur',
106     'ksp_type': 'fgmres',
107     'ksp_rtol': 1e-6,
108     'ksp_monitor_true_residual': None,
109     'pc_fieldsplit_schur': {
110         'fact_type': 'FULL',
111         'precondition': 'selfp'
112     },
```

Lines 107–109 specify the choice of iterative method (flexible GMRES) and preconditioner. Setting the `pc_type` to `fieldsplit` followed by specifying the parameter `pc_fieldsplit_type` to `schur` tells PETSc to use a block-preconditioner based on the Schur complement factorisation (5.30). We can be more specific; setting the factorisation type to be "full" (line 113) and using `selfp` as the Schur preconditioning type produces a preconditioner of the form (5.34), where the inner elliptic operator is produced by explicitly inverting a diagonal approximation to the velocity mass operator. Since our problem consists of two fields, we can further configure how each field is handled by providing additional solver parameters for each block.

```
113     'fieldsplit_0': {
114         'ksp_type': 'preonly',
115         'pc_type': 'bjacobi',
116         'sub_pc_type': 'ilu'
117     },
```

By default, the first field (velocity in our case) is labeled `fieldsplit_0`. Here, we are telling PETSc to use a single application of block-ILU to approximately invert the velocity mass operator. Now, we handle the pressure operator:

```
118     'fieldsplit_1': {
119         'ksp_type': 'cg',
120         'pc_type': 'gamg',
121         'pc_gamg_sym_graph': None,
122         'mg_levels': {
123             'ksp_type': 'chebyshev',
124             'ksp_chebyshev_esteig': None,
125             'ksp_max_it': 5,
126             'pc_type': 'bjacobi',
127             'sub_pc_type': 'ilu'
128         }
129     }
130 }
```

The pressure operator, $H$, is approximated by using a lumped diagonal approximate for the mass operator. Since the resulting approximation is elliptic and sparse, we tell PETSc to use the conjugate gradient algorithm and inner-preconditioning via PETSc's in-house multigrid implementation (GAMG). We use an aggressive set of smoothers on each multigrid level (Chebyshev accelerated block-ILU). We set the maximum number of smoother iterations on each level to be 5. Finally, we can construct our solver object with this choice of parameters:

```
up_solver = LinearVariationalSolver(up_problem,
                                    solver_parameters=params)
```

To implement the full timestepping algorithm, we need to additionally set up a solver that reconstructs $b$ after solving the eliminated problem for $\boldsymbol{u}$ and $p$. This is expressed as find $\delta b \in \mathbb{W}_b$ such that

$$\int_{\Omega} \gamma \delta b \, \mathrm{d}\boldsymbol{x} = -\frac{\Delta t}{2} N^2 \int_{\Omega} \gamma \hat{\boldsymbol{z}} \cdot \boldsymbol{u}^{(n+1)} \, \mathrm{d}\boldsymbol{x}, \quad \forall \gamma \in \mathbb{W}_b. \tag{5.38}$$

To do this, we first solve for an update and solve for the vertical velocity projected into the buoyancy space. In UFL this is expressed as follows

```
# Buoyancy solver
gamma = TestFunction(Wb)
b = TrialFunction(Wb)

a_b = gamma*b*dx
L_b = dot(gamma*khat, u0)*dx
```

We then set up a function to solve the $b$ update into, and construct a solver object for this equation too.

```
b_update = Function(Wb)
b_problem = LinearVariationalProblem(a_b, L_b, b_update)
ilu_params = {
    'ksp_type': 'cg',
    'pc_type': 'bjacobi',
    'sub_pc_type': 'ilu'
}
b_solver = LinearVariationalSolver(b_problem,
                                   solver_parameters=ilu_params)
```

This system decouples into vertical columns, and is well-conditioned (as it is a mass-matrix system, so the conditioning is independent of the grid resolution) so no special preconditioning techniques are required.

Next we organise the time loop. We set the time to zero, and initialise variables where we will put the state for the next time-step, which are initialised to the initial conditions so that we can output the resulting fields.

```
# Time-loop
t = 0
```

```
state = Function(W2 * W3 * Wb, name="state")
un1, pn1, bn1 = state.split()
un1.assign(u0)
pn1.assign(p0)
bn1.assign(b0)
output = File("results/gravity_waves.pvd")
output.write(un1, pn1, bn1)
```

Then we specify how often we want to write output and start the time loop.

```
count = 1
dumpfreq = 50
while t < tmax:
    t += dt
```

We solve the mixed system for $\boldsymbol{u}$ and $p$, and copy the result.

```
    # Solve for velocity and pressure updates
    up_solver.solve()
    un1.assign(w.sub(0))
    pn1.assign(w.sub(1))
    u0.assign(un1)
    p0.assign(pn1)
```

Then we solve the $b$ reconstruction system and copy the result.

```
    # Reconstruct buoyancy
    b_solver.solve()
    bn1.assign(assemble(b0 - 0.5*dt*N**2*b_update))
    b0.assign(bn1)
```

Finally, we output the current state every 50 time-steps.

```
    count += 1
    if count > dumpfreq:
        # Write output
        output.write(un1, pn1, bn1)
        count -= dumpfreq
```

A visualisation from a more modest resolution is provided in Figure 5.1 for the interested reader. We remark here that, on average, the linear solver converges in four Krylov iterations using the preconditioner outlined in this example. While that is respectable, more complicated systems and larger time-step sizes will require many more outer iterations for the solver to converge. In order to meet operational requirements, it is crucial to reduce the overall time-to-solution in the linear solver. An immediate way to achieve this is to reduce the number of outer iterations via using a more performant preconditioner. One possible way to achieve this is through the use of *hybridisation* techniques for mixed finite element methods.

**Fig. 5.1** The buoyancy perturbation at $t$ = 3600s. This figure was produced from a simulation with a horizontal mesh generated from 6 icosahedral refinements of a sphere and 64 vertical layers. The field is shown in a $x$-$y$ slice view with the positive $z$-direction pointing inwards.

### 5.2.3 Hybridised Mixed Methods

Hybridised finite element methods were initially introduced for the mixed formulation of second-order elliptic equations, where the formulation is replaced with a discontinuous system for the prognostic variables (Arnold and Brezzi, 1965; Brezzi and Fortin, 1991; Fraeijs De Veubeke, 1985). For mixed finite element methods, hybridisation removes the continuity of normal components of functions in $W_2$, and instead weakly imposes continuity through using Lagrange multipliers on element boundaries. In this section, we argue that hybridisation acts as an effective preconditioner, which benefits from the fact that a Schur complement factorisation of the hybridisable system can be done *exactly* (up to numerical round-off). The resulting elliptic operator is sparse, and spectrally similar to that of the Schur complement operator of the original mixed system.

We outline the hybridisation of compatible mixed method in its entirety using the model gravity wave problem. After eliminating buoyancy at the equation level (see (5.21)–(5.22)), the finite element formulation reads as follows: find $\delta \boldsymbol{u} \in W_2$ and $\delta p \in W_3$ such that

$$\int_\Omega \boldsymbol{w}\cdot\delta\boldsymbol{u}\,\mathrm{d}\boldsymbol{x} + N^2\frac{\Delta t^2}{4}\int_\Omega \boldsymbol{w}\cdot\hat{z}\delta\boldsymbol{u}\cdot\hat{z}\,\mathrm{d}\boldsymbol{x} - \frac{\Delta t}{2}\int_\Omega \nabla\cdot\boldsymbol{w}\delta p\,\mathrm{d}\boldsymbol{x} = \tilde{R}_{\boldsymbol{u}}, \tag{5.39}$$

$$\int_\Omega \phi\delta p\,\mathrm{d}\boldsymbol{x} + c^2\frac{\Delta t}{2}\int_\Omega \phi\nabla\cdot\delta\boldsymbol{u}\,\mathrm{d}\boldsymbol{x} = R_p, \tag{5.40}$$

for all $\boldsymbol{w}\in W_2$ and $\phi\in W_3$.

Recall that the main issue is the production of a dense elliptic operator due to the inversion of the dense mass matrix on the $W_2$. The main objective of using the hybridisation technique is to render the system of equations for $\boldsymbol{U}$ and $\boldsymbol{P}$ block-sparse. That is, degrees of freedom only couple within the cell, rather than across adjacent elements. Since $\boldsymbol{P}$ is already defined in a discontinuous space $W_3$, our primary target is the velocity space $W_2$.

We can immediately address the problem with inverting the velocity mass matrix by considering a larger space of functions to construct the velocity field. We introduce the "broken" function space defined as:

$$W_2^b = \{\boldsymbol{w}\in \boldsymbol{L}^2(\Omega) : \boldsymbol{w}|_K \in W_2(K), \text{ for all } K\in\Omega\}. \tag{5.41}$$

The space $W_2^b$ consists of basis functions which are locally constructed from functions in $W_2$, but the normal components are no longer continuous between elemental boundaries. Diagrams of the degrees of freedom for both the $W_2$ and $W_2^b$ are provided in Figure 5.2.

Choosing the space $W_2^b$ in our finite element discretisation has three main consequences: (1) the space is no longer a subspace of $H(\text{div})$; (2) by construction, all velocity fields terms are rendered globally discontinuous; and as a result: (3) the total number of unknowns for the velocity field is far greater than before. All interior edges/faces will have twice the number of degrees of freedom to determine (a set for the in/out-flow sides of the cell interface). We will address this point soon. An immediate positive coming from this change is that the mass operator in the velocity space, $M_{\hat{\boldsymbol{u}}} : W_2^b \to W_2^b$, given by

$$\{M_{\hat{\boldsymbol{u}}}\}_{ij} = \int_\Omega \boldsymbol{w}_i\cdot\boldsymbol{w}_j\,\mathrm{d}\boldsymbol{x} + N^2\frac{\Delta t^2}{4}\int_\Omega \boldsymbol{w}_i\cdot\hat{z}\boldsymbol{w}_j\cdot\hat{z}\,\mathrm{d}\boldsymbol{x}, \quad \boldsymbol{w}_i\in W_2^b \tag{5.42}$$

is globally block-sparse. And indeed the inverse $M_{\hat{\boldsymbol{u}}}^{-1}$ is globally sparse as a result. Furthermore, $M_{\hat{\boldsymbol{u}}}^{-1}$ can be computed directly by simply inverting the local element tensors. See Figure 5.3 for global sparsity patterns in a two-dimensional example.

Taking $\delta\hat{\boldsymbol{u}}\in W_2^b$ and testing equation (5.39) in the broken space fundamentally changes the discretisation. In particular, we have for a single cell $K$:

$$\int_K \boldsymbol{w}\cdot\delta\hat{\boldsymbol{u}}\,\mathrm{d}\boldsymbol{x} + N^2\frac{\Delta t^2}{4}\int_K \boldsymbol{w}\cdot\hat{z}\delta\hat{\boldsymbol{u}}\cdot\hat{z}\,\mathrm{d}\boldsymbol{x} - \frac{\Delta t}{2}\int_K \nabla\cdot\boldsymbol{w}\delta p\,\mathrm{d}\boldsymbol{x} + \frac{\Delta t}{2}\int_{\partial K}\boldsymbol{w}\cdot\boldsymbol{n}\delta p\,\mathrm{d}S = \tilde{R}_{\hat{\boldsymbol{u}}}, \tag{5.43}$$

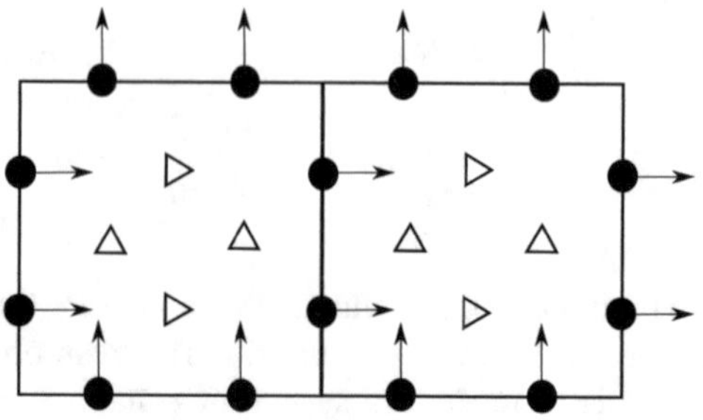

(a) $W_2$

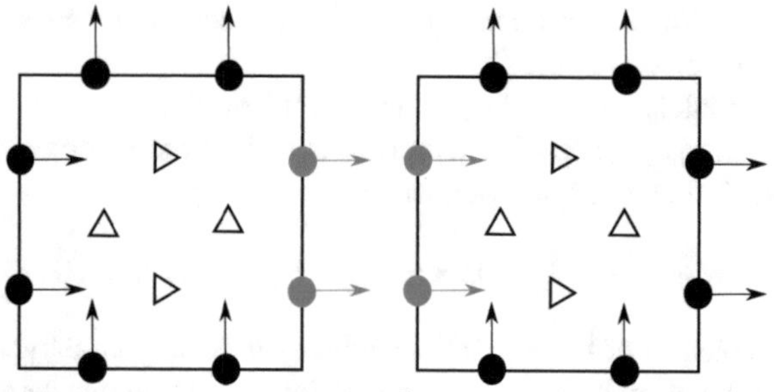

(b) $W_2^b$

**Fig. 5.2** Two-element diagrams (on quadrilaterals) for the $W_2$ space (5.2a) and the "broken" space $W_2^b$ (5.2b). These correspond to the next-to-lowest order Raviart-Thomas elements. The degrees of freedom on the edges are topologically discontinuous for the $W_2^b$ element (highlighted in blue)

(a) Mass matrix in $W_2$

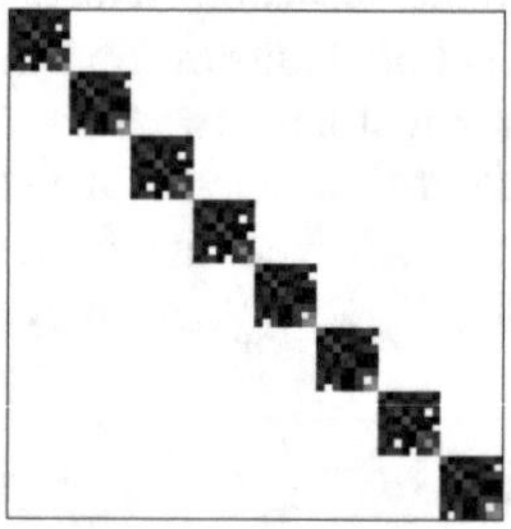

(b) Mass matrix in $W_2^b$

**Fig. 5.3** Mass matrix global sparsity patterns for the $W_2$ and $W_2^b$ operators. Note that the $W_2^b$ mass matrix is block-diagonal, similar to the pressure mass operator in $W_3$. This is also observed in three-dimensional implementations

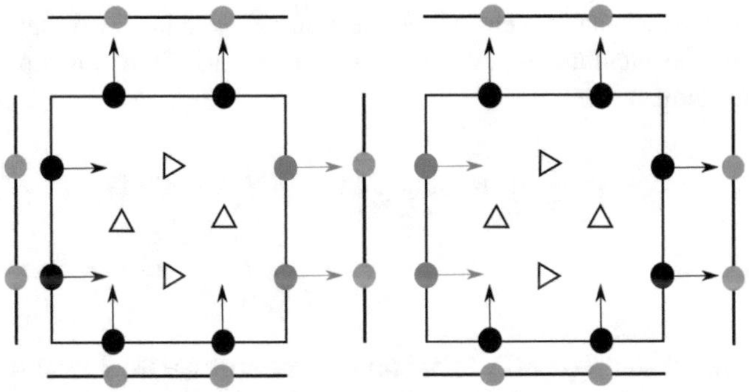

**Fig. 5.4** Two-element diagram for the space of traces of next-to-lowest order together with the broken $W_2^b$ elements. This function space is produced by taking the trace of the next-to-lowest order Raviart-Thomas space. Degrees of freedom for the traces (orange) are not mathematically defined in cell interiors, and only exist on element edges. This extends naturally to three-dimensions, with functions defined only on faces

Note here that the term $\tilde{R}_{\hat{u}}$ is identical in construction to the previous residual for the velocity. The only difference is that it consists of expressions being tested in $W_2^b$ rather than $W_2$. The surface term contains the pressure field on the boundary of the element.

Normally when testing in $W_2$, the surface integrals vanish due to the continuity of $\boldsymbol{w} \cdot \boldsymbol{n}$. That is, given two adjacent elements $K^+$ and $K^-$, the "jump" of a vector field in $W_2$ is identically zero: $\boldsymbol{w}^+ \cdot \boldsymbol{n}^+ + \boldsymbol{w}^- \cdot \boldsymbol{n}^- = 0$, where $\boldsymbol{n}^\pm$ is the outward normal on the boundaries of $K^\pm$. This is no longer the case when testing in $W_2^b$ since fields are allowed to have double-valued normal components (differing in values taken on the +/− sides).

With $\Gamma$ denoting the set of facets of the mesh, we now introduce a new function space defined only on cell interfaces. The desired space is obtained by taking the *trace* of $W_2$:

$$W_2^{\mathrm{tr}} = \{\xi \in L^2(\Gamma) : \xi|_e \in \mathcal{P}_k(e), \boldsymbol{w} \cdot \boldsymbol{n}|_e \in \mathcal{P}_k(e), \forall \boldsymbol{w} \in W_2, e \in \Gamma\}, \tag{5.44}$$

where $\mathcal{P}_k(e)$ denotes a polynomial space defined on the cell-interface $e$. Note also that for each trace function $\xi$, both $\xi|_e$ and $\boldsymbol{w} \cdot \boldsymbol{n}|_e$ belong in the *same* polynomial space. This requirement ensures consistency with the original discretisation. In general, this function space is discontinuous across vertices in two-dimensions, and vertices/edges in three-dimensions. A diagram is provided for the reader in Figure 5.4. Functions in $W_2^{\mathrm{tr}}$ are scalar-valued, with nodes on elements defined as point-evaluations at the relevant quadrature points.

Revisiting the momentum equation and testing against functions in the discontinuous space $W_2^b$, integrating the pressure gradient term in a single cell yields:

$$\int_K \boldsymbol{w}, \nabla \delta p \, \mathrm{d}\boldsymbol{x} = -\int_K \nabla \cdot \boldsymbol{w} \delta p \, \mathrm{d}\boldsymbol{x} + \int_{\partial K} \boldsymbol{w} \cdot \boldsymbol{n} \delta p \, \mathrm{d}S. \tag{5.45}$$

At this point, we introduce a new independent unknown in the trace space $W_2^{\text{tr}}$ which approximates the pressure on cell faces. This new variable $\lambda$ now appears in the *global* finite element form:

$$\int_\Omega \boldsymbol{w}\cdot\delta\hat{\boldsymbol{u}}\,\mathrm{d}\boldsymbol{x} + N^2\frac{\Delta t^2}{4}\int_\Omega \boldsymbol{w}\cdot\hat{z}\delta\hat{\boldsymbol{u}}\cdot\hat{z}\,\mathrm{d}\boldsymbol{x} - \frac{\Delta t}{2}\int_\Omega \nabla\cdot\boldsymbol{w}\delta p\,\mathrm{d}\boldsymbol{x} + \sum_{K\in\Omega}\int_{\partial K}\boldsymbol{w}\cdot\boldsymbol{n}\lambda\,\mathrm{d}S = \tilde{R}_{\hat{\boldsymbol{u}}}. \tag{5.46}$$

It can be shown that $\lambda$ physically approximates the pressure trace term $\frac{\Delta t}{2}\delta p|_\Gamma$.

We still have one final condition to address. Namely, we require that our approximation to the velocity field to respect $H(\text{div})$-conformity. In other words, we require a velocity approximation satisfying

$$\sum_{K\in\Omega}\sum_{e\in\partial K}\delta\hat{\boldsymbol{u}}\cdot\boldsymbol{n}|_e = 0. \tag{5.47}$$

We close our system by introducing the constraint equation which enforces the condition (5.47) weakly. We use test functions in $W_2^{\text{tr}}$ to act as *Lagrange multipliers* enforcing:

$$\sum_{K\in\Omega}\int_{\partial K}\xi\delta\hat{\boldsymbol{u}}\cdot\boldsymbol{n}\,\mathrm{d}S = 0, \text{ for all } \xi\in W_2^{\text{tr}}. \tag{5.48}$$

Finally, we have all the necessary components to formulate the hybridisable system entirely. The new finite element problem reads as follows: find $\delta\hat{\boldsymbol{u}}\in W_2^b$, $\delta p\in W_3$, and $\lambda\in W_2^{\text{tr}}$ such that

$$\int_\Omega \boldsymbol{w}\cdot\delta\hat{\boldsymbol{u}}\,\mathrm{d}\boldsymbol{x} + N^2\frac{\Delta t^2}{4}\int_\Omega \boldsymbol{w}\cdot\hat{z}\delta\hat{\boldsymbol{u}}\cdot\hat{z}\,\mathrm{d}\boldsymbol{x} - \frac{\Delta t}{2}\int_\Omega \nabla\cdot\boldsymbol{w}\delta p\,\mathrm{d}\boldsymbol{x} + \sum_{K\in\Omega}\int_{\partial K}\boldsymbol{w}\cdot\boldsymbol{n}\lambda\,\mathrm{d}S = \tilde{R}_{\hat{\boldsymbol{u}}}, \tag{5.49}$$

$$\int_\Omega \phi\delta p\,\mathrm{d}\boldsymbol{x} + c^2\frac{\Delta t}{2}\int_\Omega \phi\nabla\cdot\delta\hat{\boldsymbol{u}}\,\mathrm{d}\boldsymbol{x} = R_p, \tag{5.50}$$

$$\sum_{K\in\Omega}\int_{\partial K}\xi\delta\hat{\boldsymbol{u}}\cdot\boldsymbol{n}\,\mathrm{d}S = 0 = 0. \tag{5.51}$$

It can be shown that the solutions $\delta\hat{\boldsymbol{u}}$ and $\delta p$ *coincide* with the solutions $\delta\boldsymbol{u}$ and $\delta p$ of the original mixed system. That is, $\delta\hat{\boldsymbol{u}}$, albeit sought a priori in a discontinuous space, is actually $H(\text{div})$-conforming. The systems (5.39)–(5.40) and (5.49)–(5.51) are approximating the same set of equations (Arnold and Brezzi, 1985).

### 5.2.4 Discrete Hybridised System

The hybridisable system in (5.49)–(5.51) admits a block set of matrix equations. We have the usual mass operators on $W_2^b$ and $W_3$. Furthermore, we have the "broken" weak-divergence and trace-coupling operators: $\hat{D} : W_2^b \rightarrow W_3$, $C : W_2^b \rightarrow W_2^{\text{tr}}$. With $\hat{\boldsymbol{U}}$, $\boldsymbol{P}$, and $\boldsymbol{\Lambda}$ denoting the relevant coefficient vectors, the hybridisable system in matrix form is:

$$\begin{bmatrix} \widetilde{M}_{\hat{\boldsymbol{u}}} & -\frac{\Delta t}{2}\hat{D}^T & C^T \\ \frac{\Delta t}{2}c^2\hat{D} & M_p & 0 \\ C & 0 & 0 \end{bmatrix} \begin{Bmatrix} \hat{\boldsymbol{U}} \\ \boldsymbol{P} \\ \boldsymbol{\Lambda} \end{Bmatrix} = \begin{Bmatrix} \widetilde{\boldsymbol{R}_{\hat{\boldsymbol{u}}}} \\ \boldsymbol{R}_p \\ \boldsymbol{0} \end{Bmatrix}. \tag{5.52}$$

From an initial glance, this process appears to be counter-productive; the hybrid system has significantly more degrees of freedom to be determined than the original velocity-pressure system. Moreover, the full matrix operator in (5.52) is terribly conditioned and indefinite. However, the choice of function spaces provides us with essential mathematical properties to alleviate this.

Rewriting (5.52), we make the definitions:

$$\widetilde{\boldsymbol{A}} = \begin{bmatrix} \widetilde{M}_{\hat{\boldsymbol{u}}} & -\frac{\Delta t}{2}\hat{D}^T \\ \frac{\Delta t}{2}c^2\hat{D} & M_p \end{bmatrix}, \tag{5.53}$$

$$\boldsymbol{C} = \begin{bmatrix} C & 0 \end{bmatrix}, \tag{5.54}$$

$$\boldsymbol{X} = \begin{Bmatrix} \hat{\boldsymbol{U}} \\ \boldsymbol{P} \end{Bmatrix}, \tag{5.55}$$

$$\widetilde{\boldsymbol{R}} = \begin{Bmatrix} \widetilde{\boldsymbol{R}_{\hat{\boldsymbol{u}}}} \\ \boldsymbol{R}_p \end{Bmatrix}, \tag{5.56}$$

producing the block-system:

$$\widehat{\mathcal{A}} \begin{Bmatrix} \boldsymbol{X} \\ \boldsymbol{\Lambda} \end{Bmatrix} = \begin{bmatrix} \widetilde{\boldsymbol{A}} & \boldsymbol{C}^T \\ \boldsymbol{C} & 0 \end{bmatrix} \begin{Bmatrix} \boldsymbol{X} \\ \boldsymbol{\Lambda} \end{Bmatrix} = \begin{Bmatrix} \widetilde{\boldsymbol{R}} \\ \boldsymbol{0} \end{Bmatrix}. \tag{5.57}$$

By our choice of function spaces, the blocks of the operator $\widetilde{\boldsymbol{A}}$ are block-diagonal; that is, $\widetilde{\boldsymbol{A}}$ is globally discontinuous, with coupling between $\hat{\boldsymbol{U}}$ and $\boldsymbol{P}$ only restricted to cell-interiors. See Figure 5.5 for a visual comparison of the global operators. This allows us to employ a technique known as *element-wise static condensation* (Guyan, 1965).

The inverse of the Schur complement factorisation of $\widehat{\mathcal{A}}$ from (5.57) has the from:

$$\widehat{\mathcal{A}}^{-1} = \begin{bmatrix} I & -\widetilde{\boldsymbol{A}}^{-1}\boldsymbol{C}^T \\ 0 & I \end{bmatrix} \begin{bmatrix} \widetilde{\boldsymbol{A}}^{-1} & 0 \\ 0 & \boldsymbol{H}_\Gamma \end{bmatrix} \begin{bmatrix} I & 0 \\ -\boldsymbol{C}\widetilde{\boldsymbol{A}}^{-1} & I \end{bmatrix} \tag{5.58}$$

where $\boldsymbol{H}_\Gamma = -\boldsymbol{C}\widetilde{\boldsymbol{A}}^{-1}\boldsymbol{C}^T$ is the resulting reduced system for $\boldsymbol{\Lambda}$ on the mesh skeleton $\Gamma$. Unlike the Schur complement approach on the original mixed system, $\boldsymbol{H}_\Gamma$ is *sparse*,

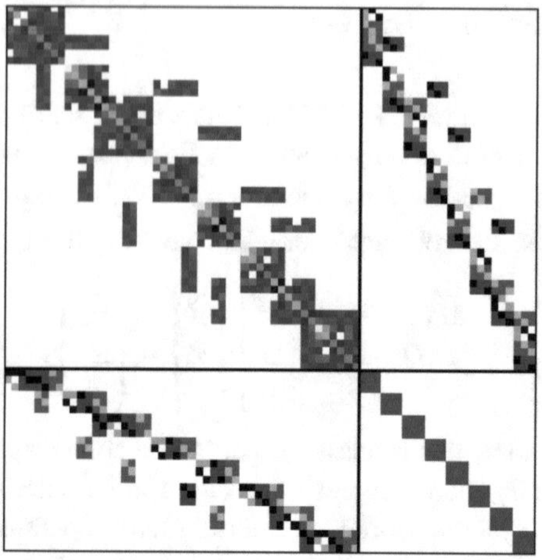

(a) Original matrix system

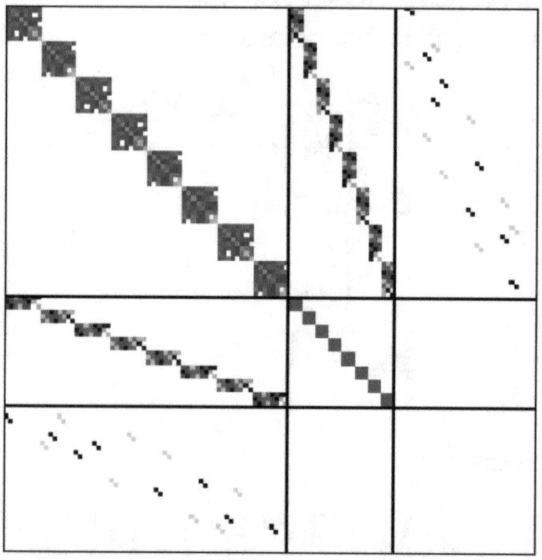

(b) Hybridised matrix system

**Fig. 5.5** Global sparsity patterns for the original discretisation 5.5a and the corresponding hybridised system 5.5b. The block-structure in 5.5b is a direct consequence of our choice in finite element spaces (rendering the velocity field discontinuous)

with a spectrum and conditioning similar to that of the pressure Helmholtz equation (Cockburn and Gopalakrishnan, 2004, 2005a, 2005b; Cockburn et al., 2009).

Solving (5.57) via a Schur complement method is equivalent to first eliminating the velocity and pressure unknowns $\boldsymbol{X}$ and solve the global problem:

$$\boldsymbol{H}_{\Gamma}\boldsymbol{\Lambda} = \boldsymbol{R}_{\Gamma} = -\boldsymbol{C}\widetilde{\boldsymbol{A}}^{-1}\widetilde{\boldsymbol{R}}. \tag{5.59}$$

Once $\boldsymbol{\Lambda}$ is determined, the prognostic variables are recovered via

$$\boldsymbol{X} = \widetilde{\boldsymbol{A}}^{-1}\left(\widetilde{\boldsymbol{R}} - \boldsymbol{C}^T\boldsymbol{\Lambda}\right). \tag{5.60}$$

Unlike with the original mixed problem, the system in (5.59) can be assembled *locally* via static condensation, without global inversion of $\widetilde{\boldsymbol{A}}$. Element-by-element, the trace operator $\boldsymbol{H}_\Gamma$ and condensed right-hand side $\boldsymbol{R}_\Gamma$ are computed by gathering the cell-wise local tensors:

$$\boldsymbol{H}_\Gamma = \{\boldsymbol{H}_{\partial K}\}_{K\in\Omega}, \quad \boldsymbol{H}_{\partial K} = -\boldsymbol{C}_K\widetilde{\boldsymbol{A}}_K^{-1}\boldsymbol{C}_K^T, \tag{5.61}$$

$$\boldsymbol{R}_\Gamma = \{\boldsymbol{R}_{\partial K}\}_{K\in\Omega}, \quad \boldsymbol{R}_{\partial K} = -\boldsymbol{C}_K\widetilde{\boldsymbol{A}}_K^{-1}\widetilde{\boldsymbol{R}}_K. \tag{5.62}$$

Similarly, $\boldsymbol{X}$ is recovered by local back-substitution; in each cell, we solve the local problem:

$$\widetilde{\boldsymbol{A}}_K\boldsymbol{X}_K = \widetilde{\boldsymbol{R}}_K - \boldsymbol{C}_K^T\boldsymbol{\Lambda}_K. \tag{5.63}$$

If $\widetilde{\boldsymbol{A}}_K^{-1}$ is computed exactly, using a suitable direct method, then solving (5.59) exactly and back-substituting (5.60) is *equivalent* to solving (5.57) via an exact Schur complement factorisation.

In practice, directly solving for $\boldsymbol{\Lambda}$ is still not practical, despite the dramatic reduction in problem size. However, we can use the fact that the operator $\boldsymbol{H}_\Gamma$ is elliptic, therefore multigrid procedures accelerated by a Krylov method can be quite effective. The construction of efficient solvers and preconditioners is a subject of on-going research and investigation. Recently, Schwarz-based preconditioners and multigrid have shown to be quite effective for a wide variety of systems arising from hybridisable finite element discretisations (Cockburn et al., 2003; Gopalakrishnan, 2009; Gopalakrishnan and Tan, 2014).

## 5.3 Hybridisation in Firedrake

In Firedrake, the process of static condensation through hybridisation is facilitated by Slate, which is a domain-specific language for describing local linear algebra i.e. linear algebra at the level of a single element (Gibson et al., 2019). The static condensation can be performed by a loop over elements, forming the local statically-condensed system involving the broken/discontinuous variables supported on that element plus the trace variables on the boundary facets of the element. This local system is then assembled into a globally coupled system that can be solved for the trace variables before solving the local system to obtain the broken/discontinuous variables.

Slate can assemble statically-condensed systems and perform reconstruction of eliminated variables for any system that can be hybridised (see Cockburn et al. 2009) for a complete description of whether systems can be hybridised or not). Further, given a linear variational solver for a hybridisable system (i.e. the system involving discontinuous field variables plus trace variables), Firedrake allows the specification through a preconditioner that static condensation should be performed, together with an interface to provide preconditioning options for the corresponding statically-condensed system. Even further, given a mixed compatible finite element system, Firedrake provides a preconditioner that takes the corresponding hybridisable system with broken velocity space, performs the static condensation, solves (possibly approximately) the condensed system, reconstructs the discontinuous variables, performs a local projection into the unbroken velocity space, and returns the result. Hence, we will only encounter Slate remotely via this preconditioner interface.

### 5.3.1 Hybridisation as a Preconditioner

The Firedrake preconditioner `HybridizationPC` takes an $H(\mathrm{div})\times L^2$ system and automatically forms the hybridisable system. This is accomplished through manipulating the UFL objects representing the discretised PDE. This includes replacing argument spaces with their discontinuous counterparts, introducing test functions on an appropriate trace space, and providing operators assembled Slate-generated kernels.

More precisely, let $\boldsymbol{Ax} = \boldsymbol{R}$ be the incoming mixed saddle point problem, where $\boldsymbol{R} = \{\boldsymbol{R}_u\ \boldsymbol{R}_p\}^T$, $\boldsymbol{x} = \{\boldsymbol{U}\ \boldsymbol{P}\}^T$, and $\boldsymbol{U}$ and $\boldsymbol{P}$ are the velocity and scalar unknowns respectively. Then this preconditioner replaces $\boldsymbol{Ax} = \boldsymbol{R}$ with the augmented system (see (5.57)):

$$\widetilde{\mathcal{A}}\begin{Bmatrix}\boldsymbol{X}\\ \boldsymbol{\Lambda}\end{Bmatrix} = \begin{Bmatrix}\widetilde{\boldsymbol{R}}\\ \boldsymbol{R}_\lambda\end{Bmatrix}, \tag{5.64}$$

where $\widetilde{\boldsymbol{R}} = \{\widetilde{\boldsymbol{R}}_{\hat{u}}\ \boldsymbol{R}_p\}^T$, $\boldsymbol{R}_\lambda$ are the right-hand sides, and $\widetilde{\cdot}$ indicates co-vectors with discontinuous functions. Here, $\boldsymbol{X} = \{\hat{\boldsymbol{U}}\ \boldsymbol{P}\}^T$ are the hybridisable (discontinuous) unknowns to be determined.

The preconditioning operator for the hybridisable system (5.64) is derived from the Schur complement factorisation in (5.58):

$$\widetilde{\mathcal{P}} = \begin{bmatrix} I & -\widetilde{\boldsymbol{A}}^{-1}\boldsymbol{C}^T \\ 0 & I\end{bmatrix}\begin{bmatrix}\widetilde{\boldsymbol{A}}^{-1} & 0\\ 0 & \boldsymbol{H}_\Gamma\end{bmatrix}\begin{bmatrix} I & 0\\ -\boldsymbol{C}\widetilde{\boldsymbol{A}}^{-1} & I\end{bmatrix}. \tag{5.65}$$

A single globally coupled system for $\Lambda$ is the only system requiring iterative inversion:

$$\boldsymbol{H}_\Gamma\boldsymbol{\Lambda} = \boldsymbol{R}_\lambda - \boldsymbol{C}\widetilde{\boldsymbol{A}}^{-1}\widetilde{\boldsymbol{R}}. \tag{5.66}$$

The recovery of $\hat{\boldsymbol{U}}$ and $\boldsymbol{P}$ happens as summarised in Section 5.2.4.

Since the velocity is constructed in a discontinuous space $W_2^b$, we must project the computed solution into $W_2$. This can be done cheaply via local facet averaging. The resulting solution is then updated via $\boldsymbol{U} \leftarrow \Pi_{\mathrm{div}}\hat{\boldsymbol{U}}$, where $\Pi_{\mathrm{div}} : W_2^b \to W_2$ is the projection mapping. This ensures the residual for the original mixed problem is properly evaluated to test for convergence. With $\widehat{\boldsymbol{P}}$ as in (5.65), the preconditioning operator for the original $\boldsymbol{A}\boldsymbol{x} = \boldsymbol{R}$ system is:

$$\mathcal{P} = \Pi \widetilde{\mathcal{P}} \Pi^T, \quad \Pi = \begin{bmatrix} \Pi_{\mathrm{div}} & 0 & 0 \\ 0 & I & 0 \end{bmatrix}. \tag{5.67}$$

In general, when Neumann conditions are present, then $\boldsymbol{R}_\lambda$ is not necessarily zero-valued. Since the preconditioner has access to the entire Python context (which includes a list of boundary conditions and the spaces in which they are applied), surface integrals on the exterior boundary are added where appropriate and incorporated in the generated Slate expressions. A more subtle issue that requires extra care is the incoming right-hand side tested in $W_2$.

The situation we are given is that we have

$$\boldsymbol{R}_{\boldsymbol{u}} = \boldsymbol{R}_{\boldsymbol{u}}(\boldsymbol{w}), \quad \boldsymbol{w} \in W_2, \tag{5.68}$$

but require $\widetilde{\boldsymbol{R}}_{\hat{\boldsymbol{u}}}(\hat{\boldsymbol{w}})$ for $\hat{\boldsymbol{w}} \in W_2^b$. For consistency, we also require for any $\boldsymbol{w} \in W_2$ that

$$\widetilde{\boldsymbol{R}}_{\hat{\boldsymbol{u}}}(\boldsymbol{w}) = \boldsymbol{R}_{\boldsymbol{u}}(\boldsymbol{w}). \tag{5.69}$$

We can construct such a $\widetilde{\boldsymbol{R}}_{\hat{\boldsymbol{u}}}$ satisfying (5.69) in the following way. By construction of the space $W_2^b$, we have for $\boldsymbol{\psi}_i \in W_2$:

$$\boldsymbol{\psi}_i = \begin{cases} \hat{\boldsymbol{\psi}}_i & \boldsymbol{\psi}_i \text{ associated with an exterior facet node,} \\ \hat{\boldsymbol{\psi}}_i^+ + \hat{\boldsymbol{\psi}}_i^- & \boldsymbol{\psi}_i \text{ associated with an interior facet node,} \\ \hat{\boldsymbol{\psi}}_i & \boldsymbol{\psi}_i \text{ associated with a cell interior node,} \end{cases} \tag{5.70}$$

where $\hat{\boldsymbol{\psi}}_i, \hat{\boldsymbol{\psi}}_i^\pm \in W_2^b$, and $\hat{\boldsymbol{\psi}}_i^\pm$ are functions corresponding to the positive and negative restrictions associated with the $i$-th facet node. In other words, these are the two "broken" parts of $\boldsymbol{\psi}_i$ on a particular facet connecting two elements. So for two adjacent cells, a basis function in $W_2$ for a particular facet node can be decomposed into two basis functions in $W_2^b$ defined on their respective sides of the facet.

We then define our "broken" right-hand side via:

$$\widetilde{\boldsymbol{R}}_{\hat{\boldsymbol{u}}}(\hat{\boldsymbol{\psi}}_i) = \frac{\boldsymbol{R}_{\boldsymbol{u}}(\boldsymbol{\psi}_i)}{N_i}, \tag{5.71}$$

where $N_i$ is the number of cells that the degree of freedom corresponding to the basis function $\boldsymbol{\psi}_i \in W_2$ touches. Using (5.70), (5.71), and the fact that $\boldsymbol{R}_{\boldsymbol{u}}$ is linear in its argument, we can verify that our construction of $\widetilde{\boldsymbol{R}}_{\hat{\boldsymbol{u}}}$ satisfies (5.69). For further implementation details, and other applications of static condensation and hybridisation, we refer the reader to Gibson et al. (2019). As we will soon see, using the preconditioner in existing Firedrake code only involves changing the solver options.

### 5.3.2 Revisiting the Gravity Wave Example

Recalling the gravity wave example in Section 5.2.2, we can adapt the simulation to utilise the hybridisation preconditioner provided by Firedrake. The only change required involves invoking the right solver parameters:

```
# Set up linear variational solver for u-p
# (After eliminating buoyancy)
W = W2 * W3
u, p = TrialFunctions(W)
w, phi = TestFunctions(W)

# radial unit vector
khat = interpolate(x/sqrt(dot(x, x)),
                   mesh.coordinates.function_space())

a_up = (dot(w, u)
        - 0.5*dt*p*div(w)
        # Appears after eliminating b
        + (0.5*dt*N)**2*dot(w, khat)*dot(u, khat)
        + phi*p + 0.5*dt*c**2*phi*div(u))*dx

L_up = (dot(w, u0)
        + 0.5*dt*dot(w, khat*b0)
        + phi*p0)*dx

bcs = [DirichletBC(W.sub(0), 0.0, "bottom"),
       DirichletBC(W.sub(0), 0.0, "top")]

w = Function(W)
up_problem = LinearVariationalProblem(a_up, L_up, w, bcs=bcs)
params = {
    'mat_type': 'matfree',
    'pc_type': 'python',
    'pc_python_type': 'firedrake.HybridizationPC',
    'ksp_type': 'fgmres',
    'ksp_monitor_true_residual': True,
    'hybridization': {
        'ksp_type': 'cg',
        'ksp_rtol': 1e-6,
        'pc_type': 'gamg',
        'pc_gamg_sym_graph': None,
        'mg_levels': {
            'ksp_type': 'chebyshev',
            'ksp_chebyshev_esteig': None,
```

```
                'ksp_max_it': 5,
                'pc_type': 'bjacobi',
                'sub_pc_type': 'ilu'
            }
        }
}
up_solver = LinearVariationalSolver(up_problem,
                                    solver_parameters=params)
```

The code defining the problem remains unchanged. We only alter parameters in line 97. Here, we are using flexible GMRES as before, but we have switched out the approximate Schur complement preconditioner for a hybridisation method, using the same preconditioned conjugate gradient method to invert the approximate pressure operator. Running the example with this set of options, one will find that the solver converges in one Krylov iteration. We summarise here the main advantages of hybridisation-based methods.

1. Since the hybridisable system is block-sparse, a reduced operator can be formed without the need for introducing auxiliary approximations (such as replacing the velocity mass matrix with a diagonal approximation);
2. The resulting condensed system for the Lagrange multipliers has far fewer degrees of freedom than the original mixed system;
3. This method requires only inverting the condensed operator, which is *sparse* and elliptic by construction;
4. By the previous point, we may recycle standard solver procedures for discrete elliptic operators for the condensed system (such as multigrid);
5. Once the Lagrange multipliers are determined, the velocity and pressure may be recovered locally by inverting (via LU) small linear systems in each cell.

As a result of item 1 and 5, a suitably tuned linear solver for the Lagrange multipliers results in GMRES convergence in a single iteration. Therefore, to gain the most performance, we can run the simulation with `ksp_type` set to `preonly`. That is, we ask the linear solver to simply apply the preconditioner only; no outer Krylov method is needed. Moreover, the elimination process to assemble the trace system is an operation-bound process. That is, little to no parallel communication is required. Together with the fact that using hybridisation removes the need for an outer Krylov method, hence reducing the total number of global reductions (such as dot-products in iterative methods), parallel communication is further reduced. This provides a great opportunity to optimise such solvers on massively parallel supercomputers. The implementation of hybridisable discretisations for more complicated geophysical systems is currently in progress.

## 5.4 Hydrostatic Pressure Solver

A very standard thing to do in geophysical simulations is to solve for a hydrostatic pressure with a vertical component of the pressure gradient that balances the gravitational force, given a pressure boundary condition at the top of the domain. In a staggered finite difference formulation, this is quite simple, as the pressure values can be solved in each column from the top down to the bottom by computing differences. In a compatible finite element formulation, this is more complicated, because it requires the solution of a system with different test and trial spaces. In this section, we will describe how this can be dealt with; it will result in a set of decoupled column systems that can also be reduced using hybridisation.

In the compressible Euler model, the velocity equation takes the form

$$\frac{\partial \boldsymbol{u}}{\partial t} + (\boldsymbol{u} \cdot \nabla)\boldsymbol{u} + 2\boldsymbol{\Omega} \times \boldsymbol{u} = -\theta \nabla \Pi + g\hat{z}, \tag{5.72}$$

where $\theta$ is the potential temperature, and $\Pi$ is the Exner pressure. Here we concentrate on the terms on the right-hand side of this equation, using the following discretisation given in Natale et al. (2016),

$$\begin{aligned} F[\boldsymbol{w}] = &\int_\Omega \nabla \cdot (\theta \boldsymbol{w}) \Pi \, \mathrm{d}\boldsymbol{x} - \int_\Gamma [\![\theta \boldsymbol{w}]\!] \{\{\Pi\}\} \, \mathrm{d}S \\ &- \int_{\partial\Omega_{top}} \theta \boldsymbol{w} \cdot \boldsymbol{n} \Pi_0 \, \mathrm{d}S - \int_\Omega g \boldsymbol{w} \cdot \hat{z} \, \mathrm{d}\boldsymbol{x}, \quad \forall \boldsymbol{w} \in W_2^0, \end{aligned} \tag{5.73}$$

where $\Pi_0$ is the required value of $\Pi$ on the upper surface $\partial\Omega_{\mathrm{top}}$, and where $W_2^0$ is the subspace of $W_2$ such that $\boldsymbol{w} \cdot \boldsymbol{n} = 0$ on the bottom surface $\partial\Omega_{\mathrm{bottom}}$. We also introduce the averaging operator $\{\{\cdot\}\}$, which is defined as:

$$\{\{\psi\}\} = \frac{1}{2}(\psi_+ + \psi_-), \tag{5.74}$$

the average of some scalar field $\psi$ over the positive (+) and negative (−) sides of a facet on the mesh skeleton $\Gamma$. The jump terms on $\Gamma$ exist because we take $\theta \in W_b$ and $\Pi \in W_3$; $\theta$ is therefore not necessarily continuous in the horizontal direction. However, the jump terms cancel on horizontal interfaces within columns since $\theta$ is always continuous in the vertical direction. Recalling the decomposition $W_2 = W_2^h \oplus W_2^v$ (and consequently $W_2^0 = W_2^h \oplus W_2^{0,v}$), we can consider the vertical part of this pair of forces by taking $\boldsymbol{w} \in W_2^{0,v}$. Hence, we can seek a hydrostatic pressure $\Pi \in W_3$ such that

$$-\int_\Omega \nabla \cdot (\theta \boldsymbol{w}) \Pi \, \mathrm{d}\boldsymbol{x} = -\int_{\partial\Omega_{top}} \theta \boldsymbol{w} \cdot \boldsymbol{n} \Pi_0 \, \mathrm{d}S - \int_\Omega g \boldsymbol{w} \cdot \hat{z} \, \mathrm{d}\boldsymbol{x}, \quad \forall \boldsymbol{w} \in W_2^{0,v}. \tag{5.75}$$

Currently this a system with different test and trial functions. As noted by Natale et al. (2016), it can be converted into an equivalent mixed system for $(\boldsymbol{v}, \Pi) \in W_2^{0,v} \times W_3$ given by

$$\int_{\Omega} (\boldsymbol{w} \cdot \boldsymbol{v} - \nabla \cdot (\theta \boldsymbol{w})\Pi) \, \mathrm{d}\boldsymbol{x} = -\int_{\partial\Omega_{top}} \theta \boldsymbol{w} \cdot \boldsymbol{n} \Pi_0 \, \mathrm{d}S - \int_{\Omega} g \boldsymbol{w} \cdot \hat{\boldsymbol{z}} \, \mathrm{d}\boldsymbol{x}, \tag{5.76}$$

$$\int_{\Omega} \phi \nabla \cdot \boldsymbol{v} \, \mathrm{d}\boldsymbol{x} = 0, \tag{5.77}$$

for all $\boldsymbol{w} \in W_2^{0,v}$ and $\phi \in W_3$. Note that the surface terms vanish, since they only contribute on vertical side facets, where $\boldsymbol{w} \cdot \boldsymbol{n} = 0$, since $\boldsymbol{w}$ is always vertical.

We can deduce that the solution $\boldsymbol{v}$ vanishes from the following. Since divergence maps from $W_2^{0,v}$ into $W_3$, the projection of $\nabla \cdot \boldsymbol{v}$ into $W_3$ is trivial, and we obtain $\nabla \cdot \boldsymbol{v} = 0$ pointwise. Further, $\boldsymbol{v}$ points in vertical direction (because it is in the vertical space $W_2^{0,v}$), and hence it satisfies $|v|_z = 0$ point-wise. Finally, the boundary condition $\boldsymbol{v} \cdot \boldsymbol{n} = 0$ implies that $|v| = 0$ on the bottom surface, and hence $|v| = 0$ everywhere. Since $\boldsymbol{v}$ vanishes, we conclude that $\Pi$ solves the hydrostatic equation (5.75).

In fact, one can symmetrise the equation by replacing Equation (5.77) with

$$\int_{\Omega} \phi \nabla \cdot (\theta \boldsymbol{v}) \mathrm{d}\boldsymbol{x} = 0, \quad \forall \phi \in W_3. \tag{5.78}$$

Natale et al. (2016) gave a uniqueness proof for $\boldsymbol{v}$, and this equation is clearly satisfied by $\boldsymbol{v} = 0$, so we deduce that $\boldsymbol{v}$ must vanish.

### 5.4.1 A Vertically-Oriented Hybridisation Preconditioner

This equation has the form of a mixed compatible finite element problem but with coupling only within columns (as might be expected for a hydrostatic equation). In fact, it is also hybridisable. Upon relaxing $\boldsymbol{v}$ to the broken vertical space $W_2^{b,v,0}$, defined as the set

$$W_2^{b,0,v} = \{\boldsymbol{v} \in \left[L^2(\Omega)\right]^n : \boldsymbol{v}|_K \in W_2^{0,v}(K), \text{ for all } K \in \Omega\}, \tag{5.79}$$

and introducing Lagrange multipliers $\lambda \in \mathrm{Trace}(W_2^v) = W_2^{\mathrm{tr},v}$ to reinforce normal continuity of $\boldsymbol{v}$, which is only supported on the horizontal facets between neighbouring elements in the same column, we obtain the coupled *hybridisable* problem: find $\boldsymbol{v} \in W_2^{b,0,v}$, $\Pi \in W_3$, and $\lambda \in W_2^{\mathrm{tr},v}$ such that

$$\int_\Omega (\boldsymbol{w}\cdot\boldsymbol{v} - \nabla\cdot(\theta\boldsymbol{w})\Pi)\,\mathrm{d}\boldsymbol{x} + \int_\Gamma [\![\boldsymbol{w}]\!]\lambda\,\mathrm{d}S = -\int_{\partial\Omega_{top}} \theta\boldsymbol{w}\cdot\boldsymbol{n}\Pi_0\,\mathrm{d}S - \int_\Omega g\boldsymbol{w}\cdot\hat{\boldsymbol{z}}\,\mathrm{d}\boldsymbol{x}, \quad (5.80)$$

$$\int_\Omega \phi\nabla\cdot(\theta\boldsymbol{v})\,\mathrm{d}\boldsymbol{x} = 0, \quad (5.81)$$

$$\int_\Gamma \mu[\![\boldsymbol{v}]\!]\,\mathrm{d}S = 0, \quad (5.82)$$

for all $\boldsymbol{w} \in W_2^{b,0,v}$, $\phi \in W_3$, and $\mu \in W_2^{\mathrm{tr},v}$. We can use static condensation to obtain a reduced symmetric positive definite system for $\lambda$ which can be efficiently and independently solved in each column. It turns out that $\lambda$ is an approximation of $\theta\Pi$ on the horizontal facets.

### 5.4.2 Firedrake Example

The example code starts as usual by importing Firedrake and defining some problem parameters.

```
from firedrake import *
nlayers = 16                      # Number of extrusion layers
R = 6.371E6/125.0                 # Scaled radius [m]: R_earth/125.0
thickness = 1.0E4                 # Thickness [m] of the domain
degree = 1                        # Degree of finite element complex
refinements = 4                   # Number of horizontal refinements
c = Constant(343.0)               # Speed of sound
N = Constant(0.01)                # Buoyancy frequency
g = Constant(9.810616)            # Accel. due to gravity (m/s^2)
N = Constant(0.01)                # Brunt-Vaisala frequency (1/s)
p_0 = Constant(1000.0*100)        # Reference pressure (Pa, not hPa)
c_p = Constant(1004.5)            # SHC of dry air at const. pressure
R_d = Constant(287.0)             # Gas constant for dry air (J/kg/K)
kappa = 2.0/7.0                   # R_d/c_p
T_eq = 300.0                      # Isothermal atmospheric temp. (K)
p_eq = 1000.0 * 100.0             # Ref surface pressure at equator
u_max = 20.0                      # Maximum amp. of zonal wind (m/s)
d = 5000.0                        # Width parameter for Theta'
lamda_c = 2.0*pi/3.0              # Longitudinal centerpoint of Theta'
phi_c = 0.0                       # Latitudinal centerpoint of Theta'
deltaTheta = 1.0                  # Maximum amplitude of Theta' (K)
L_z = 20000.0                     # Vert. wave len. of the Theta' pert.
```

Just as in the gravity wave example, we set up an atmosphere-shaped domain on a planet 125 times smaller than Earth, and construct compatible finite element spaces on it, using tensor products. First we define the extruded mesh using a cubic coordinate field:

```
# Horizontal base mesh (cubic coordinate field)
base = CubedSphereMesh(R,
                       refinement_level=refinements,
                       degree=3)

# Extruded mesh
mesh = ExtrudedMesh(base,
                    extrusion_type='radial',
                    layers=nlayers,
                    layer_height=thickness/nlayers)
```

And now we construct our spaces on extruded quadrilateral elements using the Raviart-Thomas family complex.

```
# Horizontal elements
U1 = FiniteElement('RTCF', quadrilateral, degree)
U2 = FiniteElement('DQ', quadrilateral, degree - 1)

# Vertical elements
V0 = FiniteElement('CG', interval, degree)
V1 = FiniteElement('DG', interval, degree - 1)

# HDiv element (vertical only)
W2_ele_v = HDiv(TensorProductElement(U2, V0))

# L2 element
W3_ele = TensorProductElement(U2, V1)

# Charney-Phillips element
Wt_ele = TensorProductElement(U2, V0)

# Resulting function spaces
W2v = FunctionSpace(mesh, W2_ele_v)
W3 = FunctionSpace(mesh, W3_ele)
Wtheta = FunctionSpace(mesh, Wt_ele)
```

We then set up a potential temperature field, and perform some thermodynamic calculations to set up a background potential temperature stratification plus a perturbation.

```
theta0 = Function(Wtheta)
x = SpatialCoordinate(mesh)
```

```

# Create polar coordinates:
# Since we use a CG1 field, this is constant on layers
W_Q1 = FunctionSpace(mesh, "CG", 1)
z_expr = sqrt(x[0]*x[0] + x[1]*x[1] + x[2]*x[2]) - R
z = Function(W_Q1).interpolate(z_expr)
lat_expr = asin(x[2]/sqrt(x[0]*x[0] + x[1]*x[1] + x[2]*x[2]))
lat = Function(W_Q1).interpolate(lat_expr)
lon = Function(W_Q1).interpolate(atan_2(x[1], x[0]))

# Surface temperature
G = g**2/(N**2*c_p)
Ts_expr = G
Ts_expr += (T_eq-G)*exp(
    -(u_max*N**2/(4*g*g))*u_max*(cos(2.0*lat)-1.0)
)
Ts = Function(W_Q1).interpolate(Ts_expr)

# Surface pressure
ps_expr = p_eq*exp(
    (u_max/(4.0*G*R_d))*u_max*(cos(2.0*lat)-1.0)
)*(Ts/T_eq)**(1.0/kappa)
ps = Function(W_Q1).interpolate(ps_expr)

# Background pressure
p_expr = ps*(1 + G/Ts*(exp(-N**2*z/g)-1))**(1.0/kappa)
p = Function(W_Q1).interpolate(p_expr)

# Background temperature
Tb_expr = G*(1 - exp(N**2*z/g)) + Ts*exp(N**2*z/g)
Tb = Function(W_Q1).interpolate(Tb_expr)

# Background potential temperature
thetab_expr = Tb*(p_0/p)**kappa
thetab = Function(W_Q1).interpolate(thetab_expr)
theta_b = Function(Wtheta).interpolate(thetab)
sin_tmp = sin(lat) * sin(phi_c)
cos_tmp = cos(lat) * cos(phi_c)
r = R*acos(sin_tmp + cos_tmp*cos(lon-lamda_c))
s = (d**2)/(d**2 + r**2)
theta_pert = deltaTheta*s*sin(2*pi*z/L_z)
theta0.interpolate(theta_b)
```

To define the right-hand side linear forms, we require a radial vector to set the gravitational force:

```
khat = interpolate(x/sqrt(dot(x, x)),
                   mesh.coordinates.function_space())
```

Now we build a mixed function space to define the linear variational problem, and set up test and trial functions.

```
# Calculate hydrostatic Pi
W = W2v * W3
v, pi = TrialFunctions(W)
dv, dpi = TestFunctions(W)
```

We will need to refer to the facet normal in setting the reference pressure, so we create a symbolic object for it.

```
n = FacetNormal(mesh)
```

Now we describe equations (5.76)–(5.78) using UFL, and set the boundary condition for $\boldsymbol{v}$. In this particular problem, we set the value of $\Pi_0$ at the top of the domain and enforce a no-slip condition on the vertical velocity space on the bottom. This gives us a well-posed finite element system. The resulting finite element system is written as:

```
a = (c_p*inner(v, dv) - c_p*div(dv*theta0)*pi
     + dpi*div(v*theta0))*dx
L = (-c_p*inner(dv, n)*theta0*(p/p_0)**kappa*ds_t
     - c_p*g*inner(dv, khat)*dx)

bcs = [DirichletBC(W.sub(0), 0.0, "bottom")]

w = Function(W)
piproblem = LinearVariationalProblem(a, L, w, bcs=bcs)
```

We now want to solve using the vertically-oriented hybridisation procedure outlined in Section 5.4.1. A Firedrake implementation of the vertically-oriented hybridisation preconditioner, using the Slate package (Gibson et al., 2019), can be found in (Zenodo/SCPC, 2019). We invoke this preconditioner in a similar fashion as the previous hybridisation example for the gravity wave system:

```
params = {
    'mat_type': 'matfree',
    'pc_type': 'python',
    'pc_python_type': 'scpc.VerticalHybridizationPC',
    'ksp_type': 'preonly',
```

```
119     'vert_hybridization': {
120         'ksp_type': 'cg',
121         'ksp_rtol': 1e-8,
122         'pc_type': 'gamg',
123         'pc_gamg_sym_graph': None,
124         'mg_levels': {
125             'ksp_type': 'richardson',
126             'ksp_max_it': 5,
127             'pc_type': 'bjacobi',
128             'sub_pc_type': 'ilu'
129         }
130     }
131 }
```

The choice of solver options were selected based on the properties of the condensed system. Since we formulated the mixed finite element formulation as a symmetric system (scaling by $\theta$ as in equation (5.78)), the resulting system for the trace variables is a symmetric system. We therefore use the conjugate gradient algorithm (`cg`). To precondition the trace system, we use PETSc's smooth-aggregation multigrid implementation with Richardson-ILU smoothers on the grid levels. The full configuration for the trace system (via the option prefix `vert_hybridization`) is shown in lines 119–128.

Finally, we create a function to store the result in, set up the linear solver object, and execute the hybridisation procedure to solve the system:

```
132 pisolver = LinearVariationalSolver(piproblem,
133                                    solver_parameters=params)
134
135 pisolver.solve()
```

To verify our discussion in Section 5.4, we check that norm of the computed flux variable $\boldsymbol{v}$ is approximately zero by printing the result. We split our mixed function into the $\boldsymbol{v}$ and $\Pi$ components respectively and write out the Exner pressure field to a file. Figure 5.6 displays the resulting Exner pressure field for this example.

```
136 v, exner = w.split()
137
138 # Print the norm of v: should be close to zero
139 print(sqrt(assemble(inner(v, v)*dx)))
140 File("pressure.pvd").write(exner)
```

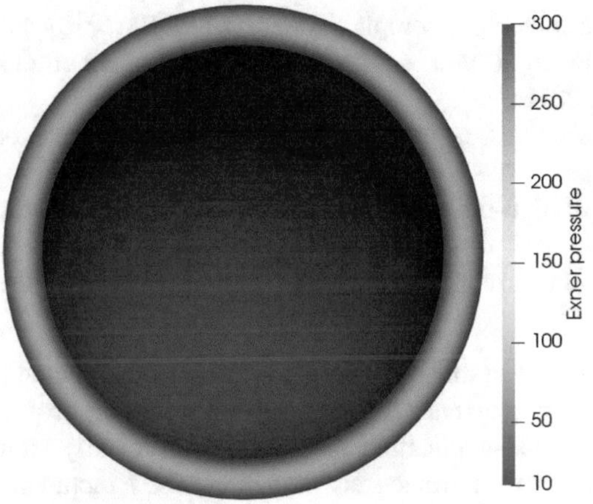


**Fig. 5.6** The Exner pressure ($y$-$z$ slice) field produced by solving equations (5.76)-(5.78). The positive $x$-direction is pointing outwards

## 5.5 Summary and Conclusions

An alternative approach to the implementation of finite element libraries is through the use of templated C++ programs. This allows the user to have full control over all the necessary components of finite element discretisations, see for example DUNE (Bastian et al., 2010). However, computational kernels (such as local assembly and application of the discrete operator) have to be written by hand. Any kernel optimisations is therefore limited by the capabilities of the available compiler. In contrast, frameworks like FEniCS (Logg and Wells, 2012; Logg et al., 2010) and Firedrake (Rathgeber et al., 2016) use domain-specific compilers to carry out optimisations that are not feasible for general purpose compilers to perform.

Expressing non-finite element operations, such as column-wise or point-wise physics parameterisations, often requires the programmer to both manage all aspects of parallelism and its incorporation into intricate numerical code. However, Firedrake provides a mechanism for the straightforward implementation of *any* localised operations as computational kernels in PyOP2 (Rathgeber et al., 2012). The PyOP2 layer is explicitly exposed to the model developer, and allows one to write and execute specialised kernels over the mesh. That is, non-finite element parameterisations and operations may still be formulated as the execution of local operations over mesh entities. Parallelisation is automatically handled by PyOP2 exactly as it is for standard Firedrake finite element kernels.

The layered set of abstractions Firedrake provides further enables discretisations that are ideal for geophysical models, equipped with sophisticated customisations:

- programmable solvers for complex systems of PDEs (Kirby and Mitchell, 2018);
- efficient kernel algorithms on layered meshes for structurally extruded grids (Bercea et al., 2016);
- construction of sophisticated finite element spaces on tensor-product grids (McRae et al., 2016);
- fast assembly of high-order finite element tensors via sum-factorisation (Homolya et al., 2017);
- and hybridisation and static condensation of finite element discretisations (Gibson et al., 2019).

The examples outlined in this book build on many of the unique features Firedrake offers for simulating geophysical flows. Several on-going projects built on top of Firedrake have enabled application specialists to efficiently simulate and explore domain areas relevant for Earth system modeling. These include:

- The Gusto Project: a compatible finite element dynamical core for three-dimensional atmospheric flows (https://firedrakeproject.org/gusto/);
- The Thetis Project: a coastal ocean model using discontinuous Galerkin methods (Kärnä et al., 2018) (https://thetisproject.org/);
- and Icepack: an ice-sheet and glacier finite element model (https://icepack.github.io/).

Together with its unique approach to software abstraction and composition, Firedrake enables the rapid implementation, development, and execution of novel numerics for complex geophysical systems.

**Code availability**

All numerical examples presented in Chaps. 3, 4, and this chapter are public and can be found at Zenodo/Springerbrief (2019). All major Firedrake components have been archived on Zenodo (2019). An installation of Firedrake with components matching those used to produce the results in this book can be obtained following the instructions at https://www.firedrakeproject.org/download.html with:

```
python3 firedrake-install --doi 10.5281/zenodo.2635069
```

# References

Adcroft A, Campin JM, Hill C, Marshall J (2004) Implementation of an atmosphere–ocean general circulation model on the expanded spherical cube. Monthly Weather Review 132(12):2845–2863, 10.1175/MWR2823.1

Alnæs MS, Logg A, Ølgaard KB, Rognes ME, Wells GN (2014) Unified Form Language: A domain-specific language for weak formulations of partial differential equations. ACM Transactions on Mathematical Software 40(2):9:1–9:37, 10.1145/2566630

Amestoy P, Duff I, L'Excellent JY (2000) Multifrontal parallel distributed symmetric and unsymmetric solvers. Computer methods in applied mechanics and engineering 184(2):501–520, 10.1016/S0045-7825(99)00242-X

Arakawa A, Lamb VR (1977) Computational design of the basic dynamical processes of the UCLA general circulation model. In: Chang J (ed) General Circulation Models of the Atmosphere, Methods in Computational Physics: Advances in Research and Applications, vol 17, Elsevier, pp 173–265, 10.1016/B978-0-12-460817-7.50009-4

Arnold DN, Brezzi F (1985) Mixed and nonconforming finite element methods: implementation, postprocessing and error estimates. ESAIM: Mathematical Modelling and Numerical Analysis 19(1):7–32, 10.1051/m2an/1985190100071

Arnold DN, Logg A (2014) Periodic table of the finite elements. SIAM News 47(9):212

Arnold DN, Falk RS, Winther R (2006) Finite element exterior calculus, homological techniques, and applications. Acta Numerica 15:1–155, 10.1017/S0962492906210018

Arnold DN, Falk RS, Winther R (2010) Finite element exterior calculus: from Hodge theory to numerical stability. Bulletin of the American Mathematical Society 47(2):281–354, 10.1090/S0273-0979-10-01278-4

 
T. H. Gibson et al., *Compatible Finite Element Methods for Geophysical Flows*, Mathematics of Planet Earth, https://doi.org/10.1007/978-3-030-23957-2

Balay S, Gropp WD, McInnes LC, Smith BF (1997) Efficient management of parallelism in object oriented numerical software libraries. In: Arge E, Bruaset AM, Langtangen HP (eds) Modern Software Tools in Scientific Computing, Birkhäuser Press, pp 163–202

Balay S, Abhyankar S, Adams MF, Brown J, Brune P, Buschelman K, Dalcin L, Dener A, Eijkhout V, Gropp WD, Kaushik D, Knepley MG, May DA, McInnes LC, Mills RT, Munson T, Rupp K, Sanan P, Smith BF, Zampini S, Zhang H, Zhang H (2018a) PETSc Web page. http://www.mcs.anl.gov/petsc, URL http://www.mcs.anl.gov/petsc

Balay S, Abhyankar S, Adams MF, Brown J, Brune P, Buschelman K, Dalcin L, Dener A, Eijkhout V, Gropp WD, Kaushik D, Knepley MG, May DA, McInnes LC, Mills RT, Munson T, Rupp K, Sanan P, Smith BF, Zampini S, Zhang H, Zhang H (2018b) PETSc users manual. Tech. Rep. ANL-95/11 - Revision 3.10, Argonne National Laboratory, URL http://www.mcs.anl.gov/petsc

Bastian P, Heimann F, Marnach S (2010) Generic implementation of finite element methods in the distributed and unified numerics environment (dune). Kybernetika 46(2):294–315

Bauer P, Thorpe A, Brunet G (2015) The quiet revolution of numerical weather prediction. Nature 525(7567):47–55, 10.1038/nature14956

Bauer W, Cotter C (2018) Energy–enstrophy conserving compatible finite element schemes for the rotating shallow water equations with slip boundary conditions. Journal of Computational Physics 373:171–187, 10.1016/j.jcp.2018.06.071

Benzi M, Golub GH, Liesen J (2005) Numerical solution of saddle point problems. Acta Numerica 14:1–137, 10.1017/S0962492904000212

Bercea GT, McRae ATT, Ham DA, Mitchell L, Rathgeber F, Nardi L, Luporini F, Kelly PHJ (2016) A structure-exploiting numbering algorithm for finite elements on extruded meshes, and its performance evaluation in Firedrake. Geoscientific Model Development 9(10):3803–3815, 10.5194/gmd-9-3803-2016

Boffi D, Brezzi F, Fortin M (2013) Mixed Finite Element Methods and Applications, Springer Series in Computational Mathematics, vol 44, 1st edn. Springer-Verlag Berlin Heidelberg, 10.1007/978-3-642-36519-5

Brezzi F, Fortin M (1991) Mixed and Hybrid Finite Element Methods, Springer Series in Computational Mathematics, vol 15, 1st edn. Springer-Verlag New York, 10.1007/978-1-4612-3172-1

Brown J, Knepley MG, May DA, McInnes LC, Smith B (2012) Composable linear solvers for multiphysics. In: Proceedings of the 2012 11th International Symposium on Parallel and Distributed Computing, IEEE Computer Society, Washington, DC, USA, ISPDC '12, pp 55–62, 10.1109/ISPDC.2012.16

Charney JG, Phillips NA (1953) Numerical integration of the quasi-geostrophic equations for barotropic and simple baroclinic flows. Journal of Meteorology 10(2):71–99, 10.1175/1520-0469(1953)010<0071:NIOTQG>2.0.CO;2

Cockburn B, Gopalakrishnan J (2004) A characterization of hybridized mixed methods for second order elliptic problems. SIAM Journal on Numerical Analysis 42(1):283–301, 10.1137/S0036142902417893

Cockburn B, Gopalakrishnan J (2005a) Error analysis of variable degree mixed methods for elliptic problems via hybridization. Mathematics of computation 74(252):1653–1677, 10.1090/S0025-5718-05-01741-2

Cockburn B, Gopalakrishnan J (2005b) New hybridization techniques. GAMM-Mitteilungen 28(2):154–182, 10.1002/gamm.201490017

Cockburn B, Gopalakrishnan J, Lazarov R (2009) Unified hybridization of discontinuous Galerkin, mixed, and continuous Galerkin methods for second order elliptic problems. SIAM Journal on Numerical Analysis 47(2):1319–1365, 10.1137/070706616

Cockburn B, Dubois O, Gopalakrishnan J, Tan S (2014) Multigrid for an HDG method. IMA Journal of Numerical Analysis 34(4):1386–1425, 10.1093/imanum/drt024

Cotter C, Kuzmin D (2016) Embedded discontinuous Galerkin transport schemes with localised limiters. Journal of Computational Physics 311:363–373, 10.1016/j.jcp.2016.02.021

Cotter C, Shipton J (2012) Mixed finite elements for numerical weather prediction. Journal of Computational Physics 231(21):7076–7091, 10.1016/j.jcp.2012.05.020

Cotter CJ, Kirby RC (2016) Mixed finite elements for global tide models. Numerische Mathematik 133(2):255–277, 10.1007/s00211-015-0748-z

Courant R, Friedrichs K, Lewy H (1928) Über die partiellen Differenzengleiehungen der mathematischen Physik. Mathematische Annalen 100(1):32–74, 10.1007/BF01448839

Crank J, Nicolson P (1947) A practical method for numerical evaluation of solutions of partial differential equations of the heat-conduction type. Mathematical Proceedings of the Cambridge Philosophical Society 43(1):50–67, 10.1017/S0305004100023197

Danilov S (2010) On utility of triangular C-grid type discretization for numerical modeling of large-scale ocean flows. Ocean Dynamics 60(6):1361–1369, 10.1007/s10236-010-0339-6

Davies T, Staniforth A, Wood N, Thuburn J (2003) Validity of anelastic and other equation sets as inferred from normal-mode analysis. Quarterly Journal of the Royal Meteorological Society 129(593):2761–2775, 10.1256/qj.02.1951

Durran DR (1989) Improving the anelastic approximation. Journal of the Atmospheric Sciences 46(11):1453–1461, 10.1175/1520-0469(1989)046, <1453:ITAA>2.0.CO;2

Durran DR (2008) A physically motivated approach for filtering acoustic waves from the equations governing compressible stratified flow. Journal of Fluid Mechanics 601:365–379, 10.1017/S0022112008000608

Fraeijs De Veubeke B (1965) Displacement and equilibrium models in the finite element method. In: Zienkiewicz OC, Holister GS (eds) Stress analysis, Wiley, New York, chap 9, pp 145–197

Gassmann A (2011) Inspection of hexagonal and triangular C-grid discretizations of the shallow water equations. Journal of Computational Physics 230(7):2706–2721, 10.1016/j.jcp.2011.01.014

Gibson TH, Mitchell L, Ham DA, Cotter CJ (2019) Slate: extending Firedrake's domain-specific abstraction to hybridized solvers for geoscience and beyond. Submitted URL https://arxiv.org/abs/1802.00303

Gopalakrishnan J (2003) A Schwarz preconditioner for a hybridized mixed method. Computational Methods in Applied Mathematics 3(1):116–134, 10.2478/cmam-2003-0009

Gopalakrishnan J, Tan S (2009) A convergent multigrid cycle for the hybridized mixed method. Numerical Linear Algebra with Applications 16(9):689–714, 10.1002/nla.636

Griffies SM, Böning C, Bryan FO, Chassignet EP, Gerdes R, Hasumi H, Hirst A, Treguier AM, Webb D (2000) Developments in ocean climate modelling. Ocean Modelling 2(3-4):123–192, 10.1016/S1463-5003(00)00014-7

Guerra JE, Ullrich PA (2016) A high-order staggered finite-element vertical discretization for non-hydrostatic atmospheric models. Geoscientific Model Development 9(5):2007–2029, 10.5194/gmd-9-2007-2016

Guyan RJ (1965) Reduction of stiffness and mass matrices. AIAA journal 3(2):380, 10.2514/3.2874

Hecht F (2012) New development in FreeFem++. Journal of Numerical Mathematics 20(3-4):251–265, 10.1515/jnum-2012-0013

Higdon RL (2006) Numerical modelling of ocean circulation. Acta Numerica 15:385–470, 10.1017/S0962492906250013

Homolya M, Kirby RC, Ham DA (2017) Exposing and exploiting structure: optimal code generation for high-order finite element methods. Submitted URL https://arxiv.org/abs/1711.02473

Homolya M, Mitchell L, Luporini F, Ham DA (2018) TSFC: A structure-preserving form compiler. SIAM Journal on Scientific Computing 40(3):C401–C428, 10.1137/17M1130642

Jackett DR, McDougall TJ (1997) A neutral density variable for the world's oceans. Journal of Physical Oceanography 27(2):237–263, 10.1175/1520-0485(1997)027 <0237:ANDVFT>2.0.CO;2

Kalnay E (2003) Atmospheric Modeling, Data Assimilation and Predictability. Cambridge University Press, 10.1017/CBO9780511802270

Kärnä T, Kramer SC, Mitchell L, Ham DA, Piggott MD, Baptista AM (2018) Thetis coastal ocean model: discontinuous Galerkin discretization for the three-dimensional hydrostatic equations. Geoscientific Model Development 11(11):4359–4382, 10.5194/gmd-11-4359-2018

Kirby RC, Mitchell L (2018) Solver composition across the PDE/linear algebra barrier. SIAM Journal on Scientific Computing 40(1):C76–C98, 10.1137/17M1133208

Kumar Das S, Weaver AJ (1995) Semi-Lagrangian advection algorithms for ocean circulation models. Journal of Atmospheric and Oceanic Technology 12(4):935–950, 10.1175/1520-0426(1995)012 <0935:SLAAFO>2.0.CO;2

Kuzmin D (2010) A vertex-based hierarchical slope limiter for *p*-adaptive discontinuous Galerkin methods. Journal of Computational and Applied Mathematics 233(12):3077–3085, 10.1016/j.cam.2009.05.028

LeVeque RJ (1996) High-resolution conservative algorithms for advection in incompressible flow. SIAM Journal on Numerical Analysis 33(2):627–665, 10.1137/0733033

Logg A, Wells GN (2010) DOLFIN: Automated finite element computing. ACM Transactions on Mathematical Software 37(2):20, 10.1145/1731022.1731030

Logg A, Mardal KA, Wells GN (eds) (2012) Automated Solution of Differential Equations by the Finite Element Method, Lecture Notes in Computational Science and Engineering, vol 84. Springer Berlin Heidelberg, Berlin, Heidelberg, 10.1007/978-3-642-23099-8

Lynch P (2008) The ENIAC forecasts: A re-creation. Bulletin of the American Meteorological Society 89(1):45–56, 10.1175/BAMS-89-1-45

McRae ATT, Cotter CJ (2014) Energy- and enstrophy-conserving schemes for the shallow-water equations, based on mimetic finite elements. Quarterly Journal of the Royal Meteorological Society 140(684):2223–2234, 10.1002/qj.2291

McRae ATT, Bercea GT, Mitchell L, Ham DA, Cotter CJ (2016) Automated generation and symbolic manipulation of tensor product finite elements. SIAM Journal on Scientific Computing 38(5):S25–S47, 10.1137/15M1021167

Natale A, Cotter CJ (2018) A variational $H(\mathrm{div})$ finite-element discretization approach for perfect incompressible fluids. IMA Journal of Numerical Analysis 38(3):1388–1419, 10.1093/imanum/drx033

Natale A, Shipton J, Cotter CJ (2016) Compatible finite element spaces for geophysical fluid dynamics. Dynamics and Statistics of the Climate System 1(1):1–31, 10.1093/climsys/dzw005

Ogura Y, Phillips NA (1962) Scale analysis of deep and shallow convection in the atmosphere. Journal of the Atmospheric Sciences 19(2):173–179, 10.1175/1520-0469(1962)019 <0173:SAODAS> 2.0.CO;2

Putman WM, Lin SJ (2007) Finite-volume transport on various cubed-sphere grids. Journal of Computational Physics 227(1):55–78, j.jcp.2007.07.022

Qaddouri A, Lee V (2011) The Canadian Global Environmental Multiscale model on the Yin-Yang grid system. Quarterly Journal of the Royal Meteorological Society 137(660):1913–1926, 10.1002/qj.873

Rathgeber F, Markall GR, Mitchell L, Loriant N, Ham DA, Bertolli C, Kelly PH (2012) Pyop2: A high-level framework for performance-portable simulations on unstructured meshes. In: 2012 SC Companion: High Performance Computing, Networking Storage and Analysis, IEEE, pp 1116–1123

Rathgeber F, Ham DA, Mitchell L, Lange M, Luporini F, McRae ATT, Bercea GT, Markall GR, Kelly PHJ (2016) Firedrake: Automating the finite element method by composing abstractions. ACM Transactions on Mathematical Software 43(3):24:1–24:27, 10.1145/2998441

Raviart PA, Thomas JM (1977) A mixed finite element method for 2nd order elliptic problems. In: Mathematical Aspects of Finite Element Methods, Springer Berlin Heidelberg, pp 292–315, 10.1007/BFb0064470

Ringler T, Thuburn J, Klemp J, Skamarock W (2010) A unified approach to energy conservation and potential vorticity dynamics for arbitrarily-structured C-grids. Journal of Computational Physics 229(9):3065–3090, 10.1016/j.jcp.2009.12.007

Rognes ME, Kirby RC, Logg A (2009) Efficient assembly of $H$(div) and $H$(curl) conforming finite elements. SIAM Journal on Scientific Computing 31(6):4130–4151, 10.1137/08073901X

Rognes ME, Ham DA, Cotter CJ, McRae ATT (2013) Automating the solution of PDEs on the sphere and other manifolds in FEniCS 1.2. Geoscientific Model Development 6(6):2099–2119, 10.5194/gmd-6-2099-2013

Shipton J, Gibson T, Cotter C (2018) Higher-order compatible finite element schemes for the nonlinear rotating shallow water equations on the sphere. Journal of Computational Physics 375:1121–1137, 10.1016/j.jcp.2018.08.027

Shu CW, Osher S (1988) Efficient implementation of essentially non-oscillatory shock-capturing schemes. Journal of Computational Physics 77(2):439–471, 10.1016/0021-9991(88)90177-5

Silvester D, Wathen A (1994) Fast iterative solution of stabilised Stokes systems Part II: Using general block preconditioners. SIAM Journal on Numerical Analysis 31(5):1352–1367, 10.1137/0731070

Skamarock WC, Klemp JB (1994) Efficiency and accuracy of the Klemp–Wilhelmson time-splitting technique. Monthly Weather Review 122(11):2623–2630, 10.1175/1520-0493(1994)122 <2623:EAAOTK>2.0.CO;2

Staniforth A, Côté J (1991) Semi-Lagrangian integration schemes for atmospheric models—a review. Monthly Weather Review 119(9):2206–2223, 10.1175/1520-0493(1991)119 <2206:SLISFA>2.0.CO;2

Staniforth A, Thuburn J (2012) Horizontal grids for global weather and climate prediction models: a review. Quarterly Journal of the Royal Meteorological Society 138(662):1–26, 10.1002/qj.958

Staniforth A, Wood N (2008) Aspects of the dynamical core of a nonhydrostatic, deep-atmosphere, unified weather and climate-prediction model. Journal of Computational Physics 227(7):3445–3464, 10.1016/j.jcp.2006.11.009

Thuburn J (2008) Numerical wave propagation on the hexagonal C-grid. Journal of Computational Physics 227(11):5836–5858, 10.1016/j.jcp.2008.02.010

Thuburn J, Cotter CJ (2012) A framework for mimetic discretization of the rotating shallow-water equations on arbitrary polygonal grids. SIAM Journal on Scientific Computing 34(3):B203–B225, 10.1137/110850293

Thuburn J, Cotter CJ, Dubos T (2014) A mimetic, semi-implicit, forward-in-time, finite volume shallow water model: comparison of hexagonal–icosahedral and cubed-sphere grids. Geoscientific Model Development 7(3):909–929, 10.5194/gmd-7-909-2014

Vallis GK (2017) Atmospheric and Oceanic Fluid Dynamics: Fundamentals and Large-Scale Circulation, 2nd edn. Cambridge University Press

Williamson DL (2007) The evolution of dynamical cores for global atmospheric models. Journal of the Meteorological Society of Japan 85B:241–269, 10.2151/jmsj.85B.241

Williamson DL, Drake JB, Hack JJ, Jakob R, Swarztrauber PN (1992) A standard test set for numerical approximations to the shallow water equations in spherical geometry. Journal of Computational Physics 102(1):211–224, 10.1016/S0021-9991(05)80016-6

Wood N, Staniforth A, White A, Allen T, Diamantakis M, Gross M, Melvin T, Smith C, Vosper S, Zerroukat M, Thuburn J (2014) An inherently mass-conserving semi-implicit semi-Lagrangian discretization of the deep-atmosphere global non-hydrostatic equations. Quarterly Journal of the Royal Meteorological Society 140(682):1505–1520, 10.1002/qj.2235

Wubs FW, de Niet AC, Dijkstra HA (2006) The performance of implicit ocean models on B- and C-grids. Journal of Computational Physics 211(1):210–228, 10.1016/j.jcp.2005.05.012

Zängl G, Reinert D, Rípodas P, Baldauf M (2015) The ICON (ICOsahedral Non-hydrostatic) modelling framework of DWD and MPI-M: Description of the non-hydrostatic dynamical core. Quarterly Journal of the Royal Meteorological Society 141(687):563–579, 10.1002/qj.2378

Zenodo/Firedrake-20190410.0 (2019) Software used in 'Firedrake Springerbrief'. 10.5281/zenodo.2635069, URL https://doi.org/10.5281/zenodo.2635069

Zenodo/SCPC (2019) SCPC: Firedrake preconditioners for static condensation and hybridization. 10.5281/zenodo.2600329

Zenodo/Springerbrief (2019) Firedrake Springerbrief Examples. 10.5281/zenodo.2633766

# Index



 
T. H. Gibson et al., *Compatible Finite Element Methods for Geophysical Flows*,
Mathematics of Planet Earth, https://doi.org/10.1007/978-3-030-23957-2

Printed in the United States
By Bookmasters